Mathematical Models for Society and Biology

Mathematical Models for Society and Biology

Edward Beltrami

State University of New York at Stony Brook

ACADEMIC PRESS

An Elsevier Science Imprint

San Diego San Francisco New York Boston
London Sydney Toronto Tokyo

Sponsoring Editor	Barbara Holland
Editorial Coordinator	Tom Singer
Production Editor	Angela Dooley
Cover Design	Dick Hannus/Hannus Design Assoc.
Composition	International Typesetting and Composition
Printer	Hamilton Printing Co.
Cover Printer	Jaguar Advanced Graphics, Inc.

This book is printed on acid-free paper. ∞

Cover image: Mike Kelly/Getty Images

Academic Press
An imprint of Elsevier Science
525 B Street, Suite 1900, San Diego, California 92101-4495, USA
http://www.academicpress.com

Academic Press
Harcourt Place, 32 Jamestown Road, London, NW1 7BY, UK
http://www.academicpress.com

Academic Press
An imprint of Elsevier Science
200 Wheeler Road, Burlington Massachusetts 01803, USA
http://www.harcourt-ap.com

Library of Congress Catalog Number: 2001096949
International Standard Book Number: 0-12-085561-5

Printed in the United States of America
01 02 03 04 05 06 HP 9 8 7 6 5 4 3 2 1

Contents

Preface

The book before you is a new upper-level undergraduate text that shows how mathematics can illuminate fascinating problems drawn from society and biology. The book assembles an unusual array of applications, many from professional journals, that have either not appeared before or that cannot be found easily in book form. Moreover the context of most chapters are current issues of real concern, and we distance ourselves from contrived "toy models" that are merely "academic" exercises. In this book mathematics follows from the problems and not the other way around, as is often the case in other works.

Mathematical modeling is viewed as an organizing principle that enables one to handle a vast and often confusing array of facts in a parsimonious manner. A model is useful when it reveals something of the underlying dynamics, providing insight into some complex process. Although models rarely replicate reality, they can serve as metaphors for what is going on in a simple and transparent manner, a bit of a caricature perhaps but informative nonetheless. The models chosen for this book all share these qualities.

Features

There are applications of interest in political science (Chapters 2 and 9); sociology (Chapter 1); economics (Chapters 7 and 9); ecology (Chapters 1, 7, and 8); public policy, in the municipalities, and management science (Chapters 3 and 4); molecular biology (Chapter 5); epidemiology (Chapter 6); and biochemistry and cell biology (Chapters 6 and 7), among other areas. No prior knowledge of any of these fields is assumed except what an interested layperson might pick up by reading the daily newspaper; any background that is needed to understand the problem is provided in the text.

A feature of the book is that several topics reappear in different guises throughout the text, thereby giving the student alternative perspectives on different facets of the same problem. A few unifying themes are woven throughout the book using fresh insights, and these result in a more cohesive presentation. This is discussed in a final section that gives an overview of the entire book and of the modeling process.

Prerequisites

As for prerequisites, I assume that a student has had the conventional training expected of a junior-level student including basic results from the multivariate calculus and matrix theory and some elementary probability theory and linear differential equations. More exotic material is explained in the text, as needed, and in the few places where relatively sophisticated tools are required I explain the results carefully and provide appropriate references to where details can be found.

Most chapters include a section that describes the relevant mathematical techniques needed later in that chapter, and there is a short appendix on conditional probability. In particular, there is a brief but fairly complete account of systems of differential equations in Chapter 6. An instructor may wish to expand on our treatment of these topics, depending on the preparation of the students.

In my opinion, the most interesting differential equation models ultimately require that solutions be obtained numerically, and it would be useful for students to have access to a simple language for computing orbits for systems of two and three nonlinear differential equations, such as those that occur in Chapters 6 through 8. In this book, all the computer generated solutions and accompanying graphics utilized MATLAB, Version 5.

Acknowledgments

A number of colleagues at Stony Brook, past and present, have influenced the development of this book by their work at the interface between mathematics and the other sciences. Indeed, there is hardly a chapter in the book that doesn't incorporate, to some extent, the inspired research of a Stony Brook scientist. These include Akira Okubo of the Marine Sciences Research Center, Ivan Chase of the Sociology Department, Jolyon Jesty of the Health Sciences Hematology Department, William Bauer of Microbiology, Larry Slobodkin of Ecology and Evolution, and Larry Bodin of the Harriman School of Public Policy.

I want to extend my sincere thanks to the following individuals for their helpful reviews of my manuscript: Jayne Ann Harder, University of Texas, Austin, Bruce Lundberg, University of Southern Colorado, Thomas Seidman, University of Maryland, Robert White, North Carolina State University, Raleigh, Daniel Zelterman, Yale University.

Crabs and Criminals

1.1. Background

A hand reaches into the still waters of the shallow lagoon and gently places a shell on the sandy bottom. We watch. A little later a tiny hermit crab scurries out of a nearby shell and takes possession of the larger one just put in. This sets off a chain reaction in which another crab moves out of its old quarters and scuttles over to the now empty shell of the previous owner. Other crabs do the same until at last some barely habitable shell is abandoned by its occupant for a better shelter, and it remains unused.

One day the president of a corporation decides to retire. After much fanfare and maneuvering within the firm, one of the vice presidents is promoted to the top job. This leaves a vacancy which, after a lapse of a few weeks, is filled by another executive whose position is now occupied by someone else in the corporate hierarchy. Some months pass and the title of the last position to be vacated is merged with some currently held job title and the chain terminates.

A lovely country home is offered by a real estate agency when the owner dies and his widow decides to move into an apartment. An upwardly mobile young professional buys it and moves his family out of the split-level they currently own after selling it to another couple of moderate income. That couple sold their modest house in a less than desirable neighborhood to an entrepreneurial fellow who decides to make some needed repairs and rent it.

What do these examples have in common? In each case a single vacancy leaves in its wake a chain of opportunities that affect several individuals. One vacancy begets another while individuals move up the social ladder. Implicit here is the assumption that individuals want or need a resource unit (shells, houses, or jobs) that is somehow better (newer, bigger, more status) or, at least, no worse than the one they already possess. There are a limited number of such resources and many applicants. As units trickle down from prestigious to commonplace, individuals move in the opposite direction to fill the opening created in the social hierarchy.

1

A chain begins when an individual dies or retires or when a housing unit is newly built or a job created. The assumption is that each resource unit is reusable when it becomes available and that the trail of vacancies comes to an end when a unit is merged, destroyed, or abandoned, or because some new individual enters the system from the outside. For example, a rickety shell is abandoned by its last resident, and no other crab in the lagoon claims it, or else a less fortunate hermit crab, one who does not currently have a shell to protect its fragile body, eagerly snatches the last shell.

A mathematical model of movement in a vacancy chain is formulated in the next section and is based on two notions common to all the examples given. The first notion is that the resource units belong to a finite number, usually small, of categories that we refer to as states and, second, that transitions take place among states whenever a vacancy is created. The crabs acquire protective shells formerly occupied by snails that have died and these snail shells come in various size categories. These are the states. Similarly, houses belong to varying price/prestige categories, while jobs in a corporate structure can be labeled by different salary/prestige classes.

Let's now consider an apparently different situation. A crime is committed and, in police jargon, the perpetrator is apprehended and brought to justice and sentenced to "serve time" in jail. Some crimes go unsolved, however, and of the criminals that get arrested only a few go to prison; most go free on probation or because charges are dropped. Moreover even if a felon is incarcerated, or is released after arrest, or even if he was never caught to begin with, it is quite possible that the same person will become a recidivist, namely, a repeat offender. What this has in common with the mobility examples given earlier is that here, too, there are transitions between states, where in this case "state" means the status of an offender as someone who has just committed a crime, or has just been arrested, or has just been jailed or, finally, has "gone straight" never to repeat a crime again. This, too, is a kind of social mobility and we will see that it fits the same mathematical framework that applies to the other examples.

One of the problems associated with models of social mobility is the difficult chore of obtaining data regarding the frequency of moves between states. If price, for example, measures the state of housing, then what dollar bracket constitutes a single state? Obviously the narrower we make a price category, the more homogeneous is the housing stock that lies within a given grouping. On the other hand, this homogeneity requires a large number of states, which exacerbates the data-gathering effort necessary to estimate the statistics of moves between states.

We chose the crab story to tell because it is a recent and well-documented study that serves as a parable for larger scale problems in sociology connected with housing and labor. It is not beset by some of the technical issues that crop up in these other areas, such as questions of race that complicate moves within the housing and labor markets. By drastically simplifying the criminal justice system we are also able to address some significant questions about the chain of moves of career criminals that curiously parallel those of crabs on the sandy sea bottom. These examples are discussed in Sections 1.3 and 1.4.

1.2. Absorbing Markov Chains

We began this chapter with examples of states that describe distinct categories such as the status of a felon in the criminal justice system or the sizes of snail shells in a lagoon. Our task now is to formalize this idea mathematically.

The behavior of individual crabs or criminals is largely unpredictable and so we consider their aggregate behavior by observing many incidents of shell swapping or by examining many crime files in public archives.

Suppose there are N states and that $p_{i,j}$ denotes the observed fraction of all moves from a given state i to all other states j. If a large number of separate moves are followed, the fraction $p_{i,j}$ represents the probability of a transition from i to j. In fact this is nothing more than the usual empirical definition of probability as a frequency of occurrence of some event. The N by N array $\mathbf{P}$ with elements $p_{i,j}$ is called a *transition matrix*.

To give an example, suppose that a particle can move among the integers $1, 2, \ldots, N$ by bouncing one step at a time either right or left. If the particle is at integer i it goes to $i + 1$ with probability p and to $i - 1$ with probability q, $p + q = 1$, except when i is either 1 or N. At these boundary points the particle stays put. It follows that the transition probabilities are given by

$$p_{i,i+1} = p \quad \text{and} \quad p_{i,i-1} = q \quad \text{for } 2 < i < N - 1$$

$$p_{1,1} = p_{N,N} = 1 \quad \text{and} \quad p_{i,j} = 0 \quad \text{for all other } j$$

The set of transitions from states i to states j is called a *random walk with absorbing barriers* and is illustrated schematically in Figure 1.1 for the case $N = 5$.

A *Markov chain* (after the Russian mathematician A. Markov) is defined to be a random process in which there is a sequence of moves between N states such that the probability of going to state j in the next step depends only on the current state i and not on the previous history of the process. Moreover, this probability does not depend on when the process is observed. The random walk example is a Markov chain since the decision to go either right or left from state i is independent of how the particle got to i in the first place, and the probabilities p and q remain the same regardless of when a move takes place.

To put this in more mathematical terms, if X_n is a random variable that describes the state of the system at the nth step then prob$(X_{n+1} = j \mid X_n = i)$, which means "the conditional probability that $X_{n+1} = j$ given that $X_n = i$" is uniquely given

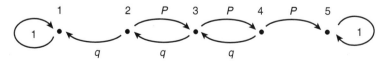

Figure 1.1. Schematic representation of a random walk.

by $p_{i,j}$. In effect, a move from i to j is statistically independent from all the moves that led to i and is also independent of the step we happen to stumble on in our observations. Clearly $p_{i,j} \geq 0$ and, since a move from state i to some other state always takes place (if one includes the possibility of remaining in i), then the sum of the elements in the ith row of the matrix $\mathbf{P}$ add to one:

$$\sum_{j=1}^{N} p_{i,j} = 1 \quad 1 \leq i \leq N$$

The extent to which these conditions for a Markov chain are actually met by crabs or criminals is discussed later. Our task now is to present the mathematics necessary to enable us to handle models of social mobility.

Let $\mathbf{P}^{(n)}$ be the matrix of probabilities $p_{i,j}^{(n)}$ of going from state i to state j in exactly n steps. This is conceptually different from the n-fold matrix product $\mathbf{P}^n = \mathbf{PP}\ldots\mathbf{P}$. Nevertheless they are equal:

Lemma 1.1 $\mathbf{P}^n = \mathbf{P}^{(n)}$

Proof: Let $n = 2$. A move from i to j in exactly two steps must pass through some intermediate state k. Because the passages from i to k and then from k to j are independent events (from the way a Markov chain was defined), the probability of going from i to j through k is the product $p_{i,k} p_{k,j}$ (Figure 1.2). There are N disjoint events, one for each intermediate state k, and so

$$p_{i,j}^{(2)} = \sum_{k=1}^{N} p_{i,k} p_{k,j}$$

which we recognize as the i, jth element of the matrix product $\mathbf{P}^2$.   □

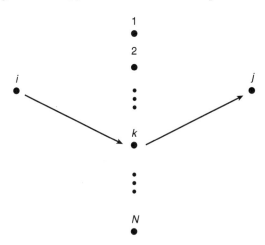

Figure 1.2. Two-step transition between states i and j through an intermediate state k.

We now proceed to the general case by induction. Assume the Lemma is true for $n - 1$. Then an identical argument shows that

$$p_{i,j}^{(n)} = \sum_{k=1}^{N} p_{i,k} p_{k,j}^{(n-1)}$$

which is the i,jth element of $\mathbf{P}^n$.

A state i is called *absorbing* if it is impossible to leave it. This means that $p_{i,i} = 1$. In the random walk example, for instance, the states 1 and N are absorbing.

Two nonabsorbing states are said to communicate if the probability of reaching one from the other in a finite number of steps is positive. Finally, an *absorbing Markov chain* is one in which the first s states are absorbing, the remaining $N - s$ nonabsorbing states all communicate, and the probability of reaching every state $i \le s$ in a finite number of steps from each $i' > s$ is positive.

It is convenient to write the transition matrix of an absorbing chain in the following block form:

$$\mathbf{P} = \begin{pmatrix} \mathbf{I} & \mathbf{0} \\ \mathbf{R} & \mathbf{Q} \end{pmatrix} \tag{1.1}$$

where $\mathbf{I}$ is an s by s identity matrix corresponding to fixed positions of the s absorbing states, $\mathbf{Q}$ is an $N - s$ by $N - s$ matrix that corresponds to moves between the nonabsorbing states, and $\mathbf{R}$ consists of transitions from transient to absorbing states. In the random walk with absorbing barriers with $N = 5$ states (Figure 1.1), for example, the transition matrix may be written as

$$
\mathbf{P} =
\begin{array}{c}
 \\ 1 \\ 5 \\ 2 \\ 3 \\ 4
\end{array}
\begin{array}{c}
\begin{array}{ccccc} 1 & 5 & 2 & 3 & 4 \end{array} \\
\begin{pmatrix}
1 & 0 & 0 & 0 & 0 \\
0 & 1 & 0 & 0 & 0 \\
q & 0 & 0 & p & 0 \\
0 & 0 & q & 0 & p \\
0 & p & 0 & q & 0
\end{pmatrix}
\end{array}
$$

Let f_i be the probability of returning to state i in a finite number of moves given that the process begins there. This is sometimes called the first return probability. We say that state i is recurrent or transient if $f_i = 1$ or $f_i < 1$, respectively. The absorbing states in an absorbing chain are clearly recurrent and all others are transient.

The number of returns to state i, including the initial sojourn in i, is denoted by N_i. This is a random variable taking on values $1, 2, \ldots$. The defining properties of a Markov chain assure that each return visit to state i is independent of previous visits and so the probability of exactly m returns is

$$\text{prob}(N_i = m) = f_i^{m-1}(1 - f_i) \tag{1.2}$$

The right side of (1.2) is known as a *geometric distribution* and it describes the probability that a first success occurs at the mth trial of a sequence of independent Bernoulli trials. In our case "success" means not returning to state i in a finite number of steps. The salient properties of the geometric distribution are discussed in most introductory probability texts and are reviewed in Exercise 1.5.1.

The probability of only a finite number of returns to state i is obtained by summing over the disjoint events $N_i = m$:

$$\text{prob}(N_i < \infty) = \sum_{m=1}^{\infty} \text{prob}(N_i = m) = \sum_{m=1}^{\infty} f_i^{m-1}(1 - f_i) = \begin{cases} 0 & \text{if } i \text{ is recurrent} \\ 1 & \text{if } i \text{ is transient} \end{cases}$$

With probability one, therefore, there are only a finite number of returns to a transient state.

In the study of Markov chains the leading question is what happens in the long run as the number of transitions increases. The next result answers this for an absorbing chain.

Lemma 1.2 *The probability of eventual absorption in an absorbing Markov chain is one.*

Proof: Each transient state can be visited only a finite number of times, as we have just seen. Therefore, after a sufficiently large number of steps, the process is trapped in an absorbing state.

The submatrix $\mathbf{Q}$ in (1.1) is destined to play an important role in what follows. We begin by recording an important property of $\mathbf{Q}$: □

Theorem 1.1 *The matrix $\mathbf{I} - \mathbf{Q}$ has an inverse.*

Proof: From Lemma 1.1 the elements of the matrix product $\mathbf{Q}^n$ represent the probability of a transition in exactly n steps from some transient state $i > s$ to some other transient $j > s$. From Lemma 1.2, $\mathbf{Q}^n$ must go to zero as n tends to infinity. □

Let $\mathbf{u}$ be an eigenvector of $\mathbf{Q}$ with corresponding eigenvalue λ. It follows immediately that $\mathbf{Q}^n\mathbf{u} = \lambda^n\mathbf{u}$. But, since $\mathbf{u}$ is fixed, the vectors $\mathbf{Q}^n\mathbf{u}$ go to zero as n increases, which implies that λ^n also goes to zero. Hence $|\lambda| < 1$. This shows that 1 is never an eigenvalue of $\mathbf{Q}$ or, to put it another way, the determinant of $\mathbf{I} - \mathbf{Q}$ is nonzero. This is equivalent to the invertibility of $\mathbf{I} - \mathbf{Q}$.

We turn next to a study of the matrix $(\mathbf{I} - \mathbf{Q})^{-1}$. Our arguments may seem to be a bit abstract but actually they are only an application of the idea of conditional probability and conditional expectation.

Let $t_{i,j}$ be the average number of times that the process finds itself in a transient state j given that it began in some transient state i. If j is different from i then $t_{i,j}$ is found by computing a conditional mean, reasoning much as in Lemma 1.1.

In fact, the passage from i to j is through some intermediate state k. Given that the process moves to k in the first step (with probability $p_{i,k}$) the mean number of times that j is visited beginning in state k is now $t_{k,j}$. The unconditional mean is therefore $p_{i,k}t_{k,j}$ and we need to sum these terms over all transient states since these correspond to disjoint events (see Figure 1.2):

$$t_{i,j} = p_{i,s+1}t_{s+1,j} + \cdots + p_{i,N}t_{N,j}$$

In the event that $i = j$, the value of $t_{i,i}$ is increased by one since the process resides in state i to begin with. Therefore, for all states i and j for which $s < i, j \leq N$,

$$t_{i,j} = \delta_{i,j} + \sum_{k=s+1}^{N} p_{i,k}t_{k,j} \tag{1.3}$$

where $\delta_{i,j}$ equals one if $i = j$, and is zero otherwise. In matrix terms (1.3) can be written as $\mathbf{T} = \mathbf{I} + \mathbf{QT}$, where $\mathbf{T}$ is the $N - s$ by $N - s$ matrix with entries $t_{i,j}$. It follows that $\mathbf{T} = (\mathbf{I} - \mathbf{Q})^{-1}$.

Let t_i be a random variable that gives the number of steps prior to absorption, starting in state i. The expected value of t_i is

$$E(t_i) = \sum_{j=s+1}^{N} t_{i,j} \tag{1.4}$$

which is the ith component of the vector $\mathbf{Tc}$, where

$$\mathbf{c} = \begin{pmatrix} 1 \\ 1 \\ \vdots \\ 1 \end{pmatrix}$$

and $\mathbf{T} = (\mathbf{I} - \mathbf{Q})^{-1}$. Vector $\mathbf{Tc}$ has $N - s$ components and the ith one can therefore be interpreted as the average number of steps before absorption when a chain begins in transient state i.

The probability $b_{i,j}$ that absorption occurs in state $j \leq s$, given that it began in some transient state i, can now be computed. Either state j is reached in a single step from i (with probability $p_{i,j}$) or there is first a transition into another transient state k and from there the process is eventually absorbed in j (with probability $b_{k,j}$). The reasoning is similar to that employed earlier. That is, since the moves from i to k and then from k to j are independent by our Markov chain assumptions, we sum over $N - s$ disjoint events corresponding to different intermediate states k:

$$b_{i,j} = p_{i,j} + \sum_{k=s+1}^{N} p_{i,k} b_{k,j} \quad s < i \le N, \; j \le s \tag{1.5}$$

In matrix terms (1.5) becomes $\mathbf{B} = \mathbf{R} + \mathbf{QB}$, where $\mathbf{R}$ and $\mathbf{Q}$ are defined in (1.1).

Now let $h_{i,j}$ be the probability that a transient state j is ever reached from another transient state i in a finite number of moves. If j differs from i then evidently

$$t_{i,j} = h_{i,j} t_{j,j}$$

and, because we must add one to the count of $t_{i,j}$ when $i = j$, in all cases one obtains

$$t_{i,j} = \delta_{i,j} + h_{i,j} t_{j,j}. \tag{1.6}$$

In matrix terms this is expressed as $\mathbf{T} = \mathbf{I} + \mathbf{HT}_{\mathrm{diag}}$, where $\mathbf{T}_{\mathrm{diag}}$ is the matrix whose only nonzero elements are the diagonal entries of $\mathbf{T} = (\mathbf{I} - \mathbf{Q})^{-1}$ and $\mathbf{H}$ is the matrix with entries $h_{i,j}$. Therefore

$$\mathbf{H} = (\mathbf{T} - \mathbf{I})\mathbf{T}_{\mathrm{diag}}^{-1}$$

Note, for later use, that $h_{i,i} = f_i$.

Relations (1.1) through (1.6) will be used in the remainder of this chapter.

1.3. Social Mobility

The tiny hermit crab *Pagarus longicarpus* does not possess a hard protective mantle to cover its body and so it is obliged to find an empty snail shell to carry around as a portable shelter. These empty refuges are scarce and only become available when its occupant dies.

In a recent study of hermit crab movements in a tidal pool off Long Island sound (see the references to Chase and others in Section 1.6) an empty shell was dropped into the water in order to initiate a chain of vacancies. This experiment was repeated many times to obtain a sample of over five hundred moves as vacancies flowed from larger to generally smaller shells. A Markov chain model was then constructed using about half this data to estimate the frequency of transitions between states, with the other half deployed to form empirical estimates of certain quantities, such as average chain length, that could be compared with the theoretical results obtained from the model itself. The complete set of experiments took place over a single season during which the conditions in the lagoon did not alter significantly. Moreover each vacancy move appeared to occur in a way that disregarded the history of previous moves. This leads us to believe that a Markov chain model is probably justifiable, a belief that is vindicated somewhat by the comparisons between theory and experiment to be given later.

There are seven states in the model. When a crab that is presently without a shelter occupies an empty shell, a vacancy chain terminates, and we label the first state to be a vacancy that is taken by a naked crab. This state is absorbing. If an empty shell is abandoned, in the sense that no crab occupies it during the 45 min of observation time, this also corresponds to an absorbing state, which we label as state 2. The remaining five states represent empty shells in different size categories, with state 3 the largest and state 7 the smallest. The largest category consists of shells weighing over 2 g, the next size class is between 1.2 and 2 g, and so on, until we reach the smallest group of shells that weigh between 0.3 and 0.7 g. Table 1.1 gives the results of 284 moves showing how a vacancy migrated from shells of size category i (namely, states $i > 2$) to shells of size j (states $j > 2$) or to an absorbing state $j = 1$ or 2. Thus, for example, a vacancy moved 9 times from a shell of the largest size (state 3) to a medium size shell in state 5, while only one of the largest shells was abandoned (absorbing state 2).

Dividing each entry in Table 1.1 by the respective row total gives an estimate for the probability of a one-step transition from state i to state j. This is displayed in Table 1.2 as a matrix in the canonical form of an absorbing Markov chain (relation 1.1).

Table 1.1. The number of moves between states in a crab vacancy chain.

From/to	1	2	3	4	5	6	7	Total moves
3	0	1	2	7	9	2	0	21
4	0	2	0	3	19	17	1	42
5	4	23	0	2	20	11	10	70
6	6	24	0	0	10	26	26	92
7	2	30	0	0	0	5	22	59

Table 1.2. Transition matrix of the crab vacancy chain.

	1	2	3	4	5	6	7
1	1	0	0	0	0	0	0
2	0	1	0	0	0	0	0
3	0	.048	.095	.333	.429	.095	0
4	0	.048	0	.071	.452	.405	.024
5	.057	.329	0	.029	.286	.157	.143
6	.065	.261	0	0	.109	.283	.283
7	.034	.508	0	0	0	.085	.373

We are now in a position to apply the theory of absorbing chains developed earlier in this chapter. The five by five submatrix in the lower right of Table 1.2 is $\mathbf{Q}$ and $(\mathbf{1} - \mathbf{Q})^{-1}$ is easily computed using any of the matrix software packages currently available, or it can be done more painfully by hand. In either case the result is

$$
\mathbf{T} = (\mathbf{I} - \mathbf{Q})^{-1} = \begin{array}{c} 3 \\ 4 \\ 5 \\ 6 \\ 7 \end{array}
\begin{pmatrix}
1.105 & .429 & 1.044 & .684 & .565 \\
0 & 1.103 & .836 & .880 & .632 \\
0 & .047 & 1.500 & .419 & .535 \\
0 & .008 & .241 & 1.541 & .753 \\
0 & 0 & .033 & .209 & 1.703
\end{pmatrix}
$$

$$
\begin{array}{ccccc} 3 & 4 & 5 & 6 & 7 \end{array}
$$

where the entries denote the average number of times $t_{i,j}$ that the process is in transient state j given that it began in transient i. Of interest to us are the components of the vector $\mathbf{Tc}$, where $\mathbf{c}$ is the vector defined previously as having all entries equal to 1. These numbers give the average number of steps required for a vacancy chain to terminate, given that it starts with an empty shell of size i. For example, an average of 3.817 moves take place before absorption whenever an empty shell of the largest size category begins a chain.

Table 1.3 compares averages computed from the model with those obtained empirically through observations and we see that there is reasonable agreement. Because no shells of size 7 were put into the water the last entry in the first column is missing.

Finally we compute the probability that a vacancy chain terminates in state 1 or 2, given that it begins with a shell of size i. Using relation (1.5), this gives us Table 1.4.

For example, starting with a fairly small shell of size 6 the probability that the last shell in a chain remains unoccupied (abandoned) is .861. This high probability reflects the fact that shells that remain at the bottom of the chain are generally cramped and in poor condition, unattractive shelters for all but the most destitute crabs.

Table 1.3. Observed and predicted lengths of crab vacancy chains.

Origin state	Observed length	Model computed length
3	3.556	3.817
4	3.323	3.443
5	2.667	2.487
6	2.567	2.538
7		1.939

Table 1.4. Probability that a crab vacancy chain ends
in a particular absorbing state.

Origin state	Absorption state 1	Absorption state 2
3	.123	.877
4	.126	.874
5	.130	.870
6	.139	.861
7	.073	.927

Each vacancy is mirrored by a crab moving to a new home except when the last shell is abandoned (absorption in state 2). In this case, the average number of crabs moving to new quarters when a vacancy chain begins in state i is one less than the average vacancy chain length. When a "naked" crab takes the last shell, on the other hand, the average number of crab moves is the same as the average chain length. Conditioning on these two events we compute the average crab mobility M_i in a chain that begins in state i. Simple considerations show that

$$M_i = b_{i,1} + \sum_{j=3}^{7} n_{i,j} - 1 \tag{1.7}$$

(Exercise 1.5.5, where it is shown that the M_i do not compare unfavorably to empirically determined crab mobility averages). The quantity (1.7) is of interest because it provides a measure of the accumulated benefit to all crabs in a lagoon resulting from a single commodity becoming available. Because crab size is closely correlated to shell size, those crabs that are able to obtain less cramped shelters tend to grow larger and produce more offspring. The impact of a single vacancy has a multiplier effect as the benefits trickle down to the community of crabs. A similar conclusion would apply in chains initiated by the opening of a new job in some organization or by the sale of a home. For example, all real estate agents benefit from the sale of a single house because this triggers a bunch of other sales, and the government also benefits by being able to collect multiple sales tax payments.

The averages $t_{i,j}$ give an estimate of the impact that the introduction of a shell of some given size will have on the mobility of crabs and therefore on their growth and reproductive capabilities. From Table 1.3 we see that placing a large shell of type 3 into the pool benefits crabs in the intermediate state 5 more than those in state 4. Evidently the smaller crabs show a preference for a larger than necessary shelter and may delay their reproductive activities until such a unit becomes available. The same conclusion could apply to shrimp, octopus, and lobsters that take shelter in rock crevices and coral reef openings. Therefore, if the goal is to improve the fitness of these animals, either as a disinterested act of conservation or as a less benign attempt to provide better fishing harvests, then a useful strategy would be to place artificial shelters such as cinder blocks in an appropriate location. The problem is to estimate the benefit that certain creatures would reap from resource units of a certain size. In the case of hermit crabs, at least, the model here suggests an answer.

1.4. Recidivism

A felon who commits another crime is said to recidivate. Because a criminal is unlikely to confess to a crime unless caught, the true probability of recidivism is unknown. To the police a *recidivist* is someone who is rearrested, whereas the correctional system regards a recidivist as someone who returns to jail. There is a need, therefore, to clarify the meaning of these crime statistics, especially because they are often reported in the press and quoted in official reports.

To gain some insight into this problem a simple Markov chain model is formulated that consists of four states describing the status of an offender as seen by the criminal justice system. The first state corresponds to a former criminal who dies or decides to "go straight" and therefore, one way or the other, does not rejoin the criminal fraternity. This state is absorbing. The remaining states correspond to an individual who has, respectively, just committed a crime, just been arrested, just been incarcerated (Figure 1.3).

Let p be the true, but unknown, probability of recidivism (a criminal, not caught, commits another crime). We assume that someone who has been arrested and then released has the same propensity p to recidivate as does someone just released from jail. This means, in effect, that the future behavior of a criminal is independent of when he or she returns to society.

From a published paper on the subject (see the references in Section 1.6) the probability p_A of being arrested, given that a crime has just been committed, is estimated to be .25 and therefore the unconditional probability of crime repetition is, in this case, $p(1 - p_A) = .75p$. Similarly the probability p_I of being convicted, sentenced, and institutionalized, given that an arrest just took place, is also estimated to be .25. Hence the unconditional probability of crime repetition, given

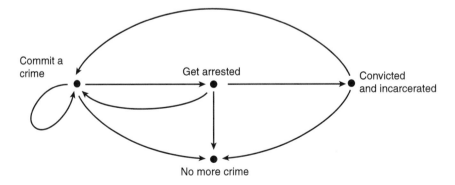

Figure 1.3. Transitions between states in a criminal justice system.

that an arrest took place, is $p(1 - p_I) = .75p$. From all transient states $i = 2, 3,$ 4 there is also a one-step absorption probability corresponding to someone returning to society as a law-abiding citizen or becoming deceased. Thus, for example, from state 2 just after a crime has been committed, the probability of no recidivism is simply $(1 - p)(1 - p_A)$. This expresses the fact that absorption into state 1 requires two independent events to hold, namely, no arrest took place after the crime was carried out and that the offender's criminal career comes to a halt.

Criminal records are prone to errors and are inherently incomplete because they do not include some arrests that may take place outside a local jurisdiction or because a central court file may fail to include arrests for minor offenses. Moreover, some individuals are falsely arrested and convicted, whereas others are dismissed from prosecution, even if arrested, because the charges are dropped due to insufficient evidence. Nevertheless, we will assume that these blemishes in the data can be disregarded and that the estimated probabilities are essentially correct. Having said this, we can write down the one-step transition matrix as

$$\mathbf{P} = \begin{array}{c} \\ 1 \\ 2 \\ 3 \\ 4 \end{array} \begin{pmatrix} 1 & 2 & 3 & 4 \\ 1 & 0 & 0 & 0 \\ .75(1-p) & .75p & .25 & 0 \\ .75(1-p) & .75p & 0 & .25 \\ 1-p & p & 0 & 0 \end{pmatrix}$$

The necessary conditions for a Markov chain model are assumed to hold. This means that a move from any state is unaffected by the past criminal history of an individual and that the transition probabilities do not change over time (which is roughly true when these numbers are estimated from data sets spanning a limited number of years).

The matrix $\mathbf{P}$ is in the canonical form (1.1) of an absorbing chain in which the block on the lower right is $\mathbf{Q}$. We can compute the 3 by 3 matrix $\mathbf{T} = (\mathbf{I} - \mathbf{Q})^{-1}$ either by hand (simple enough) or by an appropriate computer code to obtain

$$\mathbf{T} = \frac{1}{1-p} \begin{pmatrix} 1 & 1/4 & 1/16 \\ p & 1 - 3p/4 & 1/4 - 3p/16 \\ p & p/4 & 1 - 15p/16 \end{pmatrix}$$

The question of immediate interest to us is the probability of recidivism, given that an individual is in any of the transient states $i = 2, 3, 4$ or, to put it in other terms, the probability of ever returning to transient state i given that it begins there. Because $f_i = h_{i,i}$, we see from relation (1.6) that

$$t_{i,i} = f_i t_{i,i} + 1$$

where f_i is the probability of ever returning to state i in a finite number of steps.

Therefore

$$f_i = 1 - \frac{1}{t_{i,i}} \tag{1.8}$$

The i,jth component of **T** is $t_{i,j}$ and so one need only look at the diagonal components of **T** to compute (1.8). The result is

$$f_2 = p$$
$$f_3 = \frac{p}{4 - 3p}$$
$$f_4 = \frac{p}{16 - 15p}$$

That f_2 should equal p is not unexpected, because we assumed this to be true initially. Now if $p = .9$, meaning that there is a high likelihood of crime repetition, then the probability f_3 of rearrest is .69, whereas the probability of reincarceration f_4 is only .36. The different estimates of recidivism are therefore consistent with each other and simply reflect the fact separate elements of the criminal justice system (the criminal, the police, or the corrections officer) see crime repetition from different points of view.

From **T** we also see that the average number of career crimes $t_{2,2}$ is $1/(1 - p)$. When $p = .9$, the offender commits an average of 10 crimes during his lifetime, whereas with $p = .8$, there are only 5 career crimes. Thus an 11% decrease in the propensity to commit another crime can reduce the number of crimes actually carried out by 50%. This suggests that if increased vigilance on the part of the police has even a small effect in deterring a criminal this can have a substantial impact on reducing the number of crimes actually committed.

1.5. Exercises

1.5.1. A sequence of statistically independent trials each have two outcomes, success or failure, with probabilities p and $q = 1 - p$, respectively. The random variable X denotes the number of trials until a first success and one sees that $\text{prob}(X = k) = q^{k-1}p$, $k = 1, 2, \ldots$. This is called the *geometric probability distribution*. Show that

$$\sum_{k=1}^{\infty} q^{k-1} p = 1$$

and that the expected value of X is given by

$$E(X) = \sum_{k=1}^{\infty} k q^{k-1} p = \frac{1}{p}$$

In Section 1.2, $q = f_i$ and $p = 1 - f_i$. Hint: what is the sum of a geometric series?

1.5.2. Recall the definition of the random variable N_i as the number of times that a Markov chain process is in state i. Using the previous exercise show that the mean value of this variable is $E(N_i) = 1/(1 - f_i)$. Now let $i = j$ in relation (1.6) to conclude that $h_{i,I} = f_i$, a statement that also follows directly from the definitions.

1.5.3. Consider a random walk model consisting of $N + 1$ states $i = 0, 1, \ldots, N$ in which one-step transitions between states take place with the following probabilities:

$$p_{i,i+1} = p$$
$$p_{i,i-1} = 1 - p$$
$$\text{for } i = 1, 2, \ldots, N - 1 \text{ and}$$
$$p_{0,0} = p_{N,N} = 1$$

All other transitions have zero probabilities. This is an absorbing Markov chain, as we saw earlier.

In a later chapter we will have occasion to show, using a different argument than used so far, that the probability $b_{i,N}$ of absorption in state N, given that the random walk starts in transient state i, is i/N if $p = 1/2$. Verify this result using relation (1.5) in the case of $N = 4$.

1.5.4. Show that the crab mobility formula (1.7) is correct. Compare the average number of crab moves, given that a vacancy begins in state i, with the following numbers obtained empirically from observations:

Origin state	Average number of observed crab moves
3	2.61
4	2.52
5	1.75
6	1.81

1.5.5. Using the model of Section 1.4 with $p = .9$ compute the long-term probability of ever committing a crime again given that an offender has just been arrested. Hint: use relation (1.6).

Once again, let $p = .9$ and compute the average number of career jailings for those individuals that have penetrated the criminal justice system far enough to be incarcerated once.

1.5.6. In equation (1.4) we saw that the average number of steps $E(t_i)$ to absorption for a chain that begins in state i is the ith component of the vector $\mathbf{Tc}$, where all the entries of column vector $\mathbf{c}$ are equal to one and t_i is a random variable giving the number of moves required to be absorbed beginning in i.

There is an alternative derivation of this result. Starting in i, an absorption takes place either in one step, with probability

$$\sum_{j=1}^{s} p_{i,j}$$

or the first step takes us to another transient state k. In this event, the average number of steps prior to absorption is $E(t_k + 1)$ because one move already took place in going from i to k. Therefore,

$$E(t_i) = \sum_{j=1}^{s} p_{i,j} + \sum_{k=s+1}^{N} p_{i,k} E(t_k + 1) = \sum_{j=1}^{N} p_{i,j} + \sum_{k=s+1}^{N} p_{i,k} E(t_k)$$

Row sums of the transition matrix $\mathbf{P}$ add to one and so this expression can be rewritten in matrix notation as

$$\mathbf{r} = \mathbf{c} + \mathbf{Q}\mathbf{r}$$

where $\mathbf{r}$ is the vector with entries $E(t_i)$. We would like to show that $\mathbf{r}$ is the same as $\mathbf{Tc}$. But $\mathbf{Q} = \mathbf{I} - \mathbf{T}^{-1}$ and therefore

$$\mathbf{r} = \mathbf{c} + (\mathbf{I} - \mathbf{T}^{-1})\mathbf{r}$$

which verifies that indeed $\mathbf{r} = \mathbf{Tc}$.

Essentially the same argument leads to the variance of t_i. Let $\mathbf{v}$ be the vector with components $\text{Var}(t_i)$. Show that

$$\mathbf{v} = (2\mathbf{T} - \mathbf{I})\mathbf{Tc} - (\mathbf{Tc})^2$$

where $(\mathbf{Tc})^2$ is the vector whose ith component is the square of $E(t_i)$. Then use this formula to compute $\text{Var}(t_3)$ for the crab vacancy chain problem. Hint: Recall that $\text{Var}(t_i) = E(t_i^2) - E^2(t_i)$.

1.5.7. As another example of social mobility consider a university that has a fixed number of faculty on its staff in four categories or "states": instructor, assistant, associate, or full professors. Each year the teaching staff is evaluated for promotion to the next rank. Let $p_{i,j}$ be the fraction of the staff that moves from rank i to j at the beginning of the academic year, $1 \le i$, $j \le 4$. If $j = i + 1$, this indicates a promotion, whereas $j = i$ means no change in rank. If someone in rank i leaves the university he or she is replaced by an instructor. Thus, there are probabilities p_i, q_i, and h_i for which

$$p_{i,i+1} = p_i$$
$$p_{i,i} = q_i$$
$$p_{i,1} = h_i$$

and $p_{i,j} = 0$ otherwise. We assume that $p_i + q_i + h_i = 1$.

Explain why this is a Markov chain and write down the transition matrix **P**. When is it an absorbing chain?

1.6. Further Readings

A comprehensive treatment of vacancy chains in sociology is in the book [52], whereas the specific model of crab mobility discussed in Section 1.3 is taken from the papers [19] and [51] by Chase and others.

While a thorough discussion of recidivism can be found in the book [37], the model of Section 1.4 comes from the paper [13] by Blumstein and Larson.

An excellent text on Markov chains, including absorbing chains, with many applications is in Kemeny and Snell's book [32]. By the same authors is a classic modeling text [33] with applications in the social sciences. Here, too, one finds several Markov chain examples.

It Isn't Fair: Municipal Workers, Congressional Seats, and the Talmud

2.1. Background

Most public services are required to meet a weekly demand that varies over time in an uneven manner, often over all seven days of the week and 24 hours a day. The personnel available to match this demand profile must of necessity have overlapping shifts since no individual expects to work continuously over the week. This is unlike the private sector in which work requirements are generally uniform with a fixed "nine-to-five" workday that coincides exactly with the availability of personnel.

Examples involving public services are not hard to provide. It suffices to mention sanitation workers, police officers, ambulance drivers, nurses in a hospital, security guards, emergency work crews, airport baggage handlers, and transit workers, to name a few. In all these examples, the assignment of workers to different periods of the week is complicated by the fact that required workload does not match up to the available workforce.

In New York City, for example, there is effectively no refuse collection on Sunday if we exclude certain private carting services. This means that there is even more garbage than usual waiting to be picked up on Monday morning because there is Saturday's backlog to contend with in addition to the trash generated on Sunday. More sanitation workers are needed on Monday and because some refuse, with its attendant smells and hazards, remains uncollected at the end of the day because of the overload, on Tuesday as well. One solution, of course, is to hire additional workers to fill the gap. However, in times of fiscal restraint this costly solution is unattractive to the municipality, which would prefer, instead, a restructuring of worker schedules to better match the workload. This requires some care, however. Any schedule that consists of irregular and inconvenient work shifts is unsatisfactory to labor. They prefer a regular pattern of workdays that gives the employee as many weekends off as possible, for example, or that meets some other

"days off" requirement. There is a trade-off here between the needs of the municipality, which would like to get the job done in the face of severe fiscal deficits, and
the labor union, which insists on a work schedule that is fair to its members (the
question of salary and benefits is a separate issue that is ignored here).

In the next section, we discuss a mathematical framework for worker scheduling
in the context of an actual labor dispute between the sanitation workers' union and
the city of New York that took place some years ago.

Let's switch, temporarily, to what appears to be a different situation. The U.S.
Constitution stipulates that "Representatives and direct taxes shall be apportioned
among the several states which may be included in this Union, according to their
respective numbers . . ." (Article 1, Section 2). What this says, in effect, is that a
fixed number of seats in Congress are to be allocated among the different states
so that each state has representation in proportion to its population. This idea of
one man–one vote is difficult to meet in practice because the division of seats
by population is usually a fraction that must be rounded off. Because political
power is rooted in representation, the rounding problem has been a source of
controversy and debate throughout the history of the United States ever since
Hamilton, Jefferson, and Adams first struggled to resolve this issue. They, and
their successors, attempted to devise a method that would be fair to all states,
meaning, among other things, that if a state gains in population after a new census
count, then it should not give up a seat to any state that has lost population. We will
examine this problem more closely in Section 2.3 where we see that what at first
glance appears to be a reasonable solution can turn out to violate certain obvious
criteria of fairness.

The apportionment problem should remind one of scheduling, in which a desig-
nated workload is to be shared by a fixed number of employees so that each worker
gets a satisfactory arrangement of days off. Later, in Chapter 3, an analogous question arises in the context of deploying a fixed number of emergency vehicles, such
as fire engines, to different sectors of the city so that they can respond effectively
to calls for service in a way that is equitable in terms of the workload sustained
by each fire company and, at the same time, provides coverage that is fair to all
citizens.

In a similar vein is the question of reapportionment in which a given state with a
fixed number of representatives to Congress must now decide to divide up the state
into political districts. There are a number of ways that a geographical area can
be partitioned into sectors of roughly equal population but partisan rivalry tends
to encourage "gerrymandering" in which irregularly shaped districts are formed
to favor the election of one candidate over another. In this case, the allocation
problem is compromised by considerations that are difficult to quantify.

There is an unexpected connection to the question of fair allocation of scarce
resources to problems posed in the Talmud concerning the division of an estate
among heirs whose claims exceed the available amount. Its most concise expression is framed by an ancient dispute and its resolution in the Talmud as "Two
hold a garment; one claims it all, the other claims half. Then the one is awarded

three quarters, the other one quarter." Although the meaning of fairness is arguably different in this case, it bears a superficial resemblance to dividing any limited commodity, as when urban districts need an assignment of fire companies, or states vie for congressional seats, and the fire units or seats, as the case may be, are not available in the quantities desired. One requirement for fairness in the case of the Talmud will be that any settlement among a group of claimants continues to be applicable when it is restricted to any subset of contenders. This problem is discussed in Section 2.4.

Evidently some of these questions are nearly intractable because they are too suffused with political harangue and backroom deals, while others are more susceptible to rational argument. The terms allocation, apportionment, assignment, deployment, or even districting and partitioning take on similar meanings, depending on the application, and when a nonpartisan mathematical approach can be sensibly used, it will often come in the form of an optimization problem with integer constraints. This is an approach that is common to a wide number of situations that are only superficially dissimilar, as we will see in this chapter and the next.

2.2. Manpower Scheduling

Before discussing the specific case of New York City sanitation workers, let us put the topic of scheduling workers in perspective by looking at another example first. Assume, as is the case in a number of public services, that each employee is assigned to five consecutive days of work followed by two days off. Moreover, if there are several shifts a day, a worker reports to the same time period, such as the 8:00 A.M. to 4:00 P.M. shift. This may not always be true (many urban police departments have rotating 8-hour shifts to cover a 24-hour day) but we limit ourselves to this simple situation because it already illustrates the basic ideas involved in scheduling.

Suppose that the total workforce is broken into N groups of very nearly equal size. In our case the only possibilities which exist for days off are Monday–Tuesday or Tuesday–Wednesday or . . . Sunday–Monday. Call these feasible periods "recreation schedules" and label them consecutively by the index $j = 1, 2, \ldots, 7$. Suppose that it has already been determined that in order to meet the average demand a total of n_i groups must be working on the ith day. That is, $r_i = N - n_i$ groups are permitted off that day. Now let x_j be the number of groups that will be allowed to adopt the jth recreation schedule during any given week. For example, x_3 is the number of times that Wednesday–Thursday is chosen. We would like to develop a schedule so that each group has an identical pattern of days off during an N-week horizon, on a rotating basis. What this means will become more clear as we proceed.

It is apparent that because Tuesday, for example, belongs to the first and second recreation schedules, then $x_1 + x_2$ must equal r_2. In general, the following system of constraining relations must be satisfied each week:

$$
\begin{aligned}
x_1 & & + x_7 &= r_1 \\
x_1 + x_2 & & &= r_2 \\
x_2 + x_3 & & &= r_3 \\
x_3 + x_4 & & &= r_4 \\
x_4 + x_5 & & &= r_5 \\
x_5 + x_6 & & &= r_6 \\
x_6 + x_7 &= r_7 &
\end{aligned}
\tag{2.1}
$$

This system of equations may be expressed more compactly in matrix notation as $\mathbf{Ax} = \mathbf{r}$, where $\mathbf{A}$ is the matrix given by

$$
\begin{pmatrix}
1 & 0 & 0 & 0 & 0 & 0 & 1 \\
1 & 1 & 0 & 0 & 0 & 0 & 0 \\
0 & 1 & 1 & 0 & 0 & 0 & 0 \\
0 & 0 & 1 & 1 & 0 & 0 & 0 \\
0 & 0 & 0 & 1 & 1 & 0 & 0 \\
0 & 0 & 0 & 0 & 1 & 1 & 0 \\
0 & 0 & 0 & 0 & 0 & 1 & 1
\end{pmatrix}
$$

Equations (2.1) are readily solvable by elimination. For example, if $r_1 = 1$, $r_2 = r_5 = 2, r_3 = r_4 = 3, r_6 = 5$, and $r_7 = 6$, the unique solution is $x_1 = x_5 = 0, x_2 = x_4 = 2, x_3 = x_7 = 1, x_6 = 5$. Thus the Tuesday–Wednesday schedule is followed twice, and so forth. Note that since each group gets exactly two days off a week, with N groups one must have

$$
\sum_{i=1}^{7} r_i = 2N
$$

Therefore, by adding up the Equations (2.1), it follows that

$$
\sum_{j=1}^{7} x_j = \frac{1}{2} \sum_{i=1}^{7} r_i = N
$$

This means that we can construct a rotating schedule for each group over exactly N weeks. In the example just given in which $N = 11$, an 11-week rotating schedule

Table 2.1. An 11-week rotating schedule.

	Mon	Tues	Wed	Thurs	Fri	Sat	Sun
1		★	★				
2						★	★
3		★	★				
4						★	★
5			★	★			
6						★	★
7				★	★		
8						★	★
9				★	★		
10						★	★
11	★						★

is exhibited in Table 2.1, in which ★ denotes a day off. This chart can be read as either the arrangement of days off for any given group over an 11-week period (each row represents a week) or as a snapshot in any given week of the days in which the 11 different groups are off (each row is then one of the groups). When viewed as an 11-week schedule it is understood that in the 12th week, the schedule repeats itself starting again from the first row. Also observe that it is possible to permute the rows in any fashion. The particular arrangement of Table 2.1 is designed to avoid monotony for the workers by offering a variety of days off. The important factor is that on day i exactly n_i units are working, which is independent of row permutation.

It could happen that the r_i values are different in some other shift, say 4:00 P.M. to midnight. In this case, a separate solution is obtained for this period using the same basic argument.

It would appear, therefore, that the matter has been brought to a close. However there is a hitch. Although (2.1) always possess a unique solution, it may not be an acceptable one. It is essential that the values of x_j be nonnegative and integers, but this is not at all guaranteed. Consider, in fact, the situation in which all $r_i = 1$ except for $r_5 = 2$. Then x_3 turns out to have the value $1/2$ as is easily verified.

It appears that our choice of recreation schedules was too restrictive. To see this, let us augment the set of possible recreation schedules by allowing a single day off in a week. Index Monday only off, Tuesday only off, ..., by $j = 8, ..., 14$ and let x_j denote the number of groups that are permitted the jth recreation schedule, as before. Then because Monday, for example, appears on the first, seventh, and eighth recreation schedules, one must have

$$x_1 + x_7 + x_8 = r_1$$

and, in general, the system (2.1) is replaced by

$$
\begin{aligned}
x_1 & & +x_7+x_8 & & &= r_1 \\
x_1+x_2 & & +x_9 & & &= r_2 \\
x_2+x_3 & & +x_{10} & & &= r_3 \\
& \vdots & & & & \vdots \\
x_6+x_7 & & +x_{14} & & &= r_7
\end{aligned}
\tag{2.2}
$$

It is now apparent that a nonnegative integer solution to (2.2) always exists. It suffices to set x_i to zero for $i = 1, 2, \ldots, 7$ and then let $x_{i+7} = r_i$. Of course, this is an unsatisfactory solution for the workers who now have only one day off each week. A more acceptable solution would be one in which the number of six-day workweeks is as small as possible. This leads to the problem of minimizing the sum

$$
\sum_{j=8}^{14} x_j
\tag{2.3}
$$

or, equivalently, of maximizing

$$
\sum_{j=1}^{7} x_j
$$

subject to the constraints (2.2) and the condition that the x_j be nonnegative integers. This is the first, but not the last, example that we will encounter in this book of a type of optimization problem known as an *integer program*. Generally these problems are difficult to solve, but in the present case one can take advantage of the special structure of (2.2) to develop a simple algorithm that is suitable for hand computation. We sidestep the details (see, however, Exercise 2.5.1) since our main interest here is another scheduling question.

For many years the sanitation workers of New York City worked according to a six-week rotating schedule in which everyone got Sunday off and one-sixth of the workforce had off on Monday, another sixth on Tuesday, and so forth. This meant that, Sunday aside, only 5/6th of the workers were available on any given day. But, by the decade of the seventies there was an escalating demand for refuse collection in the city, which translated into a requirement that actually 14/15th of the pickup crews should have been on the streets on Monday, and 9/10th on Tuesday, clearly a mismatch between the supply of, and the need for, workers (Figure 2.1). This discrepancy led the city to negotiate a new labor contract that would solve the problem without having to hire new workers. The story of how this came to pass

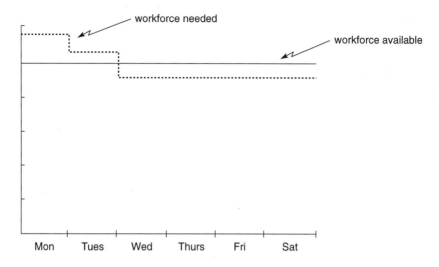

Figure 2.1. The mismatch between available workers and the demand for refuse pickups.

is told elsewhere (see the references in Section 2.6), but for us the interesting part is how a revision of the work schedule would accomplish the goal.

A proposal was made to change the existing rotating schedule by dividing the workforce into 30 rather than 6 equal groups. Recall that 14/15 of the sanitation crews are needed on Monday. This means that 28 out of the $N = 30$ groups should be working on that day or, using our previous notation, $n_1 = 28$ and $r_1 = N - n_1 = 2$. The remaining days of the week have r values that had been determined to be $r_2 = 3, r_3 = 4, r_4 = 7, r_5 = 7, r_6 = 7$, and, of course, on Sunday $r_7 = 30$.

Guided by the need for integer solutions and mindful of the example worked out earlier in this section, the admissible recreation schedules will include six-day workweeks (only Sunday off) as well as two- and three-day weekends. Let x_i denote the number of times that a Sunday–Monday, Sunday–Tuesday,..., Sunday–Saturday schedule is chosen, $i = 1, \ldots, 6$. A Friday–Saturday–Sunday three-day schedule is chosen x_7 times while a Sunday–Monday–Tuesday combination occurs x_8 times. Finally, Sunday only is scheduled x_9 times.

Labor practices and union regulations stipulate that each worker is to get an average of two days off a week. With $N = r_7 = 30$, this means that

$$\frac{1}{30} \sum_{i=1}^{7} r_i = 2$$

Because Monday appears on the first and eighth recreation schedules, it follows as before that $x_1 + x_8 = r_1$. Similar relations hold for the other days and we therefore obtain

$$
\begin{aligned}
x_1 & & & & & +x_8 &= r_1 \\
& x_2 & & & & +x_8 &= r_2 \\
& & x_3 & & & &= r_3 \\
& & & x_4 & & &= r_4 \\
& & & & x_5 + x_7 & &= r_5 \\
& & & & x_6 + x_7 & &= r_6 \\
x_1 + x_2 + \cdots & & & & & x_9 &= r_7
\end{aligned}
\tag{2.4}
$$

Adding Equations (2.4) shows that

$$
2\sum_{j=1}^{8} x_j + x_7 + x_8 + x_9 = \sum_{j=1}^{7} r_j = 2N
$$

From the last row of 2.4 and the fact that $r_7 = N = 30$ we obtain

$$
\sum_{j=1}^{9} x_j = 30
$$

Taking the last two relations together gives

$$
x_7 + x_8 = x_9
\tag{2.5}
$$

and, therefore, there are just as many six-day workweeks (Sunday only off) as there are weeks with three-day weekends off. At this point we pause to observe that in the original six-week rotating schedule a Saturday–Sunday–Monday combination would occur naturally every sixth week. The labor union would, of course, prefer to maintain this or some similarly advantageous weekend break under any revised schedule that might be proposed.

Let's begin more modestly by asking for the schedule that gives the largest number of two-day weekends. That is, we want to maximize $x_1 + x_6$, or equivalently, in view of relations (2.4) to maximize $r_1 + r_6 - x_7 - x_8$. The solution is quite apparent, namely, to let x_7 and x_8 be zero. This gives $x_i = r_i$ for $i = 1, 2, \ldots, 6$ with the remaining x_i all zero. The result is then a schedule that provides nine two-day weekends off every 30 weeks. Not bad! The rotating schedule itself is displayed in Table 2.2 and it is to be read in the same way as Table 2.1. By suitably permuting the rows in Table 2.2 one can arrange to have five two-day and two three-day weekends (Exercise 2.5.2), the schedule that was ultimately chosen in the labor negotiations for a new contract with the city of New York.

Returning now to the goal of having as many three-day weekends as possible we see immediately from (2.5) that there is a price to pay for this luxury because

Table 2.2. A 30-week rotating schedule.

	Mon	Tues	Wed	Thurs	Fri	Sat	Sun
1						★	★
2		★					★
3		★					★
4		★					★
5	★						★
6			★				★
7			★				★
8	★						★
9			★				★
10			★				★
11						★	★
12				★			★
13				★			★
14						★	★
15				★			★
16				★			★
17				★			★
18						★	★
19				★			★
20				★			★
21						★	★
22					★		★
23					★		★
24						★	★
25					★		★
26					★		★
27					★		★
28						★	★
29					★		★
30					★		★

it would require an equal number of compensatory six-day workweeks. This was not too attractive to the city since it meant that overtime pay had to be budgeted for the extra day. A compromise solution would have been to trade off two- and three-day weekends and, after much discussion, such a compromise was indeed selected, namely, the one indicated in Exercise 2.5.3.

A more formal approach to choosing a trade-off between two- and three-day weekends can be based on a decision to judge two-day weekends to be "worth"

2/3 of a three-day weekend. The goal, in this case, is to maximize the weighted sum

$$\frac{2}{3}(x_1 + x_6) + x_7 + x_8 \tag{2.6}$$

By relation (2.4) this is equivalent to maximizing

$$r_1 + r_6 - \frac{1}{3}(x_1 + x_6)$$

The solution to this problem is left to Exercise 2.5.3, where, as it turns out, only three-day weekends are selected.

We can generalize from these examples to place the scheduling problem in a more abstract setting. Suppose that there is a list of M feasible recreation schedules and that the jth schedule is chosen x_j times. Let $a_{i,j} = 1$ if the jth schedule contains day i and set $a_{i,j} = 0$ otherwise. $\mathbf{A}$ is the matrix with entries $a_{i,j}$ and we require that $\mathbf{Ax} = \mathbf{r}$, where $\mathbf{r}$ is a vector with components r_i, the given number of groups that can be allowed off on day i, and $\mathbf{x}$ is the vector with components x_j. The problem is to choose the nonnegative integers x_j so as to maximize a weighted sum

$$\sum_{j=1}^{M} c_j x_j$$

in which c_j are given nonnegative quantities. The two cases considered previously are instances of this integer programming formulation.

2.3. Apportionment

Throughout its rich history the number of states in our country has increased from the original 13 to the present 50, so that at any moment there were N states, with populations $p_1, p_2, \ldots, p_N$. The number of seats in the House of Representatives also grew from an original 65 and at a given moment in history it had a value of h, with h never less than N. The seats are apportioned among the states, with a_i allocated to state i. The sum of the a_i equals h and, of course, the a_i must be integers, with at least one per state. The exact share of h to which the i state is entitled, its quota q_i, is simply the fraction of h that is proportional to the population of that state, namely,

$$q_i = \frac{p_i h}{\sum_{i=1}^{N} p_i}$$

We have already mentioned the dilemma faced by Congress in allotting these quotas: they need not be integers!

A number of schemes have been devised over the years to mitigate this difficulty. First note that if some q_i is less than one then it must be set to unity. Because the House size remains at h, the remaining q_i must be adjusted accordingly. This leads us to define a fair share s_i as the largest of the numbers 1 and cq_i, where c is chosen so that the sum of the s_i is now h. For example, if $h = 10$ and $N = 4$, with populations of 580, 268, 102, and 50 (thousand) then the quotas q_i are 5.80, 2.68, 1.02, and 0.5. The last state is entitled to one seat and the remaining 9 seats are apportioned among the remaining three states whose quotas are now 5.49, 2.54, and 0.97. Again, the last state is given exactly one seat leaving 8 to distribute among the other two, whose revised quotas are 5.47 and 2.53. Therefore $s_1 = 5.47$, $s_2 = 2.53$, $s_3 = s_4 = 1$. In 1792 Alexander Hamilton proposed to give the i state the integer part of s_i, denoted by $[s_i]$, and to allocate the remaining

$$h - \sum_{i=1}^{N} [s_i] = k$$

seats, one each, to the states having the largest remainders. From a mathematical point of view, this amounts to choosing integers $a_i \geq 1$ so as to minimize the sum

$$\sum_{i=1}^{N} (a_i - s_i)^2 \qquad (2.7)$$

subject to the condition that

$$\sum_{i=1}^{N} a_i = h \qquad (2.8)$$

which, once again, is an integer programming problem. In effect it asks for integer allocations a_2 that are never less than unity and are as close as possible to the fair shares s_i (which need not be integers) in the sense that they minimize the squared difference (2.7) while observing the constraint (2.8). This problem admits a simple solution that is identical, as we will see, to Hamilton's proposal. Begin by assigning one seat, the least possible, to each state. Clearly $1 = a_i \leq s_i$. Next, add one more seat to the state i for which the difference

$$(a_i - s_i)^2 - (a_i + 1 - s_i)^2 = 2(s_i - a_i) - 1 \qquad (2.9)$$

is largest. The key idea here is that a_i appears only in the ith term of the sum 2.7 and so 2.7 is made smallest by allocating the additional unit to the term that most decreases the sum. Continuing in this fashion we eventually must assign $a_i = [s_i]$ seats to each state. The reason for this is that, as long as $a_i + 1 < s_i$, adding another unit to state i can only contribute to the reduction of 2.7 and this remains true until a_i is equal to $[s_i]$. Thereafter the remaining k seats are distributed, one each at most, to those states for which, by virtue of 2.9, the remainder $s_i - [s_i]$ is largest.

Hamilton's idea is, at first sight, a reasonable one and so it was adopted for a time during the last century. Eventually, however, it was shown to be flawed. This happened after the 1880 census, when the House size increased from 299 to 300. The state of Alabama had a quota of 7.646 at 299 seats and 7.671 at 300 seats, whereas Texas and Illinois increased their quotas from 9.640 and 18.640 to 9.682 and 18.702, respectively. Alabama had previously been given 8 seats of the original 299 but an application of Hamilton's method to $h = 300$ gave Texas and Illinois each an additional representative. Because only one new seat was added, this meant that Alabama was forced to give up one representative for a new total of 7. This paradoxical situation occurred in other cases as well and it resulted, ultimately, in scrapping Hamilton's idea. What was being violated here is the notion of House monotonicity, which stipulates that, in a fair apportionment, no state should lose a representative if the House size increases.

Thomas Jefferson had a different idea in 1792. He recommended that a_i be the largest of 1 and $[p_i/x]$, where x is a positive number chosen so that the sum of the a_i equal h. As before, [.] means "the greatest integer part of." The role of x is to be a unique divisor of the population p_i of each state so that representation is determined by the quotient p_i/x. If one state has a greater population than another, its share of House seats is no less than that of the smaller state.

Jefferson's method can be formulated differently. Let S be the subset of all states for which $a_i > 1$. Then, for each i in S,

$$1 < a_i = \frac{p_i}{x} - y_i$$

where $0 \le y_i < 1$ is a number that is zero when p_i/x is an integer. By cross-multiplying it follows that

$$\frac{p_i}{a_i} - \frac{x}{a_i} < x = \frac{p_i}{a_i} - \frac{xy_i}{a_i} \le \frac{p_i}{a_i}$$

That is, x does not exceed the minimum of p_i/a_i over all i in S. From the left side of the inequality we obtain $x > p_i/(a_i + 1)$. When $[p_i/x] \le 1$ then $a_i = 1$ and so $p_i/x < a_i + 1$. In all cases, then, x is greater than the largest of $p_i/(a_i + 1)$. Putting all this together gives

$$\max_{\text{all } i} \frac{p_i}{a_i + 1} < x \le \min_{i \text{ in } S} \frac{p_i}{a_i} \qquad (2.10)$$

where min and max are shorthand, respectively, for "minimum of" and "maximum of."

Conversely, if (2.10) is satisfied, then the right side of the inequality shows that $1 < a_i \le p_i/x$ for each i in S, while the left side of (2.10) gives us $a_i + 1 > p_i/x$ for all i. Therefore $a_i = [p_i/x]$ for all i in S; otherwise $a_i + 1 \le p_i/x$, which is a contradiction.

However, when Jefferson's method was applied to the 1792 census in which $h = 105$ an anomaly occurred in which Virginia's fair share of 18.310 was rewarded with 19 seats, whereas Delaware's share of 1.613 gave it only one seat. The larger state was favored over the smaller one, an imbalance that was frequently observed using Jefferson's method from 1792 to 1840 during which there were five censuses. John Quincy Adams proposed a modification of the procedure in which [.] now means "rounding up" to the next largest integer. But this tended to favor smaller states at the expense of the larger ones. Daniel Webster suggested a compromise in 1832 in which [.] was to mean that one rounds off to the nearest integer. By an argument similar to the one given above, Webster's method implies the existence of an x for which

$$\max \frac{p_i}{a_i + 1/2} \le x \le \min \frac{p_i}{a_i - 1/2} \qquad (2.11)$$

(Exercise 2.5.4) and it seemed to give results more reasonable than either of the methods proposed by Jefferson or Adams. Indeed, as Balinski and Young argue in their thorough discussion of the problem (their book is referenced in Section 2.6), Webster's method goes a long way towards satisfying a bunch of fairness criteria including the previously mentioned House monotonicity requirement. Interestingly, Webster's approach, which was adopted for only a decade beginning in 1840, also satisfies an integer program. To show this we observe first that the per capita representation of state i is a_i/p_i, whereas the ideal per capita representation, across all states of the Union, is h/p, where p is the combined population of all states together. Consider now the sum of the squared differences of a_i/p_i to h/p, weighted by the population of state i:

$$\sum_{i=1}^{N} p_i \left(\frac{a_i}{p_i} - \frac{h}{p} \right)^2 = \sum_{i=1}^{N} \frac{a_i^2}{p_i} - \frac{h^2}{p} \qquad (2.12)$$

Choosing integers $a_i \ge 1$ to minimize (2.12) subject, of course, to their sum being equal to h, we get Webster's method. In fact, since h^2/p is constant we see from (2.12) that it suffices to minimize the sum of the a_i^2/p_i. If an optimal choice has been made, then interchanging a single seat between two states r and s cannot reduce (2.12) when $a_r > 1$. Keeping the allocations to all other states the same, the interchange implies that

$$\frac{(a_r - 1)^2}{p_r} + \frac{(a_s + 1)^2}{p_s} \ge \frac{a_r^2}{p_r} + \frac{a_s^2}{p_s}$$

or,

$$\frac{p_r}{a_r - 1/2} \geq \frac{p_s}{(a_s + 1/2)} \tag{2.13}$$

We claim that

$$\min \frac{p_i}{a_i - 1/2} \geq \max \frac{p_s}{a_s + 1/2}$$

for otherwise there exist integers r and s for which (2.13) is violated. Condition (2.11) is therefore satisfied.

Although mathematical convenience often dictates the choice of an objective function in a minimization problem, the previous examples should dispel the idea that this can be done with impunity. The decision of what function to optimize must be examined carefully in terms of the intended application. Another example of this ambiguity will appear in the next chapter.

We close by touching on the reapportionment problem mentioned at the beginning of this chapter, in which a state with k representatives must now partition its geographical area into political districts. Even more than the apportionment issue, the question of districting is rooted in a struggle for power since it directly affects the ability of a candidate to be elected to Congress. The ethnic and racial makeup of a neighborhood can favor one political party over another and the way district boundaries are drawn affects the balance of votes. Nevertheless, one can at least pose some aspect of the problem mathematically and, even if it is somewhat artificial, the formulation is useful in terms of applications to more benign problems of partitioning a region into service areas such as school districts, postal zones, or police precincts.

Suppose that the state is divided into N parcels of land, census tracts to be specific, and that we wish to form k contiguous districts from these parcels to correspond to k seats in the House of Representatives. Imagine, by looking at a map of the state, that a number M of clusters of the N parcels have been formed, generally overlapping, that constitute potential political districts. The procedure for actually carrying out this partitioning is a separate matter that can itself be formalized, but we bypass this step here much as we skipped the question of how feasible "recreation schedules" were formed in the manpower scheduling problem. The only requirement in forming the M clusters is that they be feasible in terms of being reasonably compact in size and not too misshapen (not too "gerrymandered") as well as connected (no enclaves).

Let $a_{i,j} = 1$, if tract i belongs to the jth feasible cluster, and $a_{i,j} = 0$ otherwise, and let $x_j = 1$, if the jth cluster is to be actually chosen as one of the political districts, with $x_j = 0$ otherwise. The population of tract i is p_i and so the total population of cluster j is

$$p(j) = \sum_{i=1}^{N} a_{i,j} p_i$$

If p is the state's total population, then the population of each district should ideally be p/k to achieve equal representation. In effect, $p(j)$ should differ from p/k as little as possible. This leads to the integer program of minimizing the sum of squared differences

$$\sum_{j=1}^{M} x_j \left[\frac{p(j) - p}{k} \right]^2$$

subject to the constraint

$$\sum_{j=1}^{M} x_j = k$$

which ensures that exactly k districts are formed out of the M potential ones. Whether this formulation has any merit in practice is arguable but at least it poses the problem in a coherent and parsimonious manner and helps us to recognize the possible connections to similar districting problems.

2.4. An Inheritance

The Babylonian Talmud of nearly two millennia ago records religious and legal decisions of that time. The Talmud includes the Mishna, or rulings, followed by rabbinical commentaries on the Mishna. One fascinating problem recorded there tells of three heirs who lay claim to an inheritance of 100, 200, and 300 units (which, for simplicity, we think of as dollars.) However, the estate is worth less than the sum claimed by the heirs. For an estate of 100, 200, and 300 dollars, the Mishna arbitrates the claims as shown in Table 2.3.

Table 2.3.

		Claim		
		100	200	300
	100	$33\frac{1}{3}$	$33\frac{1}{3}$	$33\frac{1}{3}$
Estate	200	50	75	75
	300	50	100	150

This is a puzzling division. When the estate is small it is divided equally among the heirs but as the estate grows it is either divided proportionally or in some manner that seemingly bears no relation to the other divisions. What could they have been thinking? Is this a fair division?

To understand the rationale behind the allocations, let's quote another, more straightforward Mishna that reads "two men hold a garment; one claims it all, the other claims half. The one is awarded 3/4, the other 1/4" The principle here seems simple enough. One claims half the garment and so concedes the remaining half to the other. What continues to be in dispute is the remaining half and this is divided equally. By implication, if both claimed the full garment, it would be divided equally between them, each getting half.

A more general version of the heir's problem is for n heirs or creditors, as the case may be, having claims in the amounts of $d_1 \le d_2 \le \cdots \le d_n$ with $D = \sum d_i \ge E$, where E is the size of the estate or debt. The goal is to find a corresponding division of E so that the creditors receive shares $x_1 \le x_2 \le \cdots \le x_n$ in which $E = \sum x_i$. This satisfies at least one fairness criterion in that a larger claimant never receives a smaller amount than a lesser claimant; that is, $x_i \le x_j$, when $d_i \le d_j$.

Let's begin with a more careful analysis of the contested garment problem. Let u_+ denote the larger of u and zero. If person i imposes a lien of d_i, then the uncontested amount is $E - (E - d_1)_+ - (E - d_2)_+$ and this, as with the garment is divided equally. In addition, person i concedes an amount $(E - d_i)_+$ to person j. Thus

$$x_i = \frac{1}{2}(E - (E - d_1)_+ - (E - d_2)_+) + (E - d_j)_+ \tag{2.14}$$

for $i, j = 1, 2$ and $i \ne j$.

It is useful to restate (2.14) in the following manner to show how it depends on E. When E is less than or equal to d_1, the estate is divided equally, but as soon as E exceeds d_1 the lesser creditor receives half his or her share and then temporarily drops out while the larger claimant continues to gain an amount $E - d_1/2$ until E has reached the value of $D/2$. At this juncture $x_2 = d_2/2$, and each creditor has now achieved half their respective claims. As E continues to increase they again divide the additional amount equally until the first creditor receives his full share of $x_1 = d_1$ and departs. The balance then all goes to the higher claimant until, finally, E reaches a value of D. At this point, $x_2 = d_2$, and both creditors are vindicated.

Note that (2.14) is monotonely increasing with E so that no creditor receives less than before if E happens to increase.

To extend the two person allocation to n people another fairness criterion is invoked, which stipulates that the division is internally consistent. What this means is that the amount allocated to any subset of claimants is the same allocation they would have gotten as part of the larger group. Put another way, if S is any subset of the creditors, then the share E_s that they receive as a group from the arbitration is itself divided among themselves according to the same rule. The individual shares they each get from E_s are the same as the amounts awarded under the rule applied to the entire group of n people. In particular, if S consists of two individuals i and j, and if they receive a total between them of $x_i + x_j$ from the arbitration of n people, then the amount $E_s = x_i + x_j$ is allocated so that i gets x_i and j gets x_j. It is left as to Exercise 2.5.7 to show that a consistent arbitration rule is uniquely determined.

We limit the following discussion to the case of three people for simplicity, but a similar reasoning applies in general. A rule is proposed that is consistent with the contested garment principle when applied to a subset of two individuals.

If E is less than or equal to d_1, the total amount available is demanded by each claimant and so it is divided equally. However, as soon as E exceeds d_1 the two-person contested garment rule is applied with person 1 versus a coalition of persons 2 and 3. Thus x_1 is $d_1/2$ and the remainder is divided between 2 and 3 again according to the garment rule. This means that if E is less than or equal to d_2 the remaining amount $(E - d_1/2)/2$ is divided equally between these two individuals, whereas if E exceeds d_2, then x_2 equals $d_2/2$ and $x_3 = E - (d_1 + d_2)/2$. When E reaches $D/2$, it is easy to see that x_3 becomes $d_3/2$, and each contestant has achieved half their due. As E continues to increase the additional amount is divided equally until the lowest claimant drops out, having attained d_1. The same rule applies to the other two individuals until the one attains d_2 and the remainder goes to the highest claimant who finally achieves d_3 when E has the value D.

Applying this rule to the Table 2.3 shows that an estate of 100 is divided equally, whereas an estate of 200 ensures that the lowest claimant obtains half his lien, namely, 50, with the remainder split equally among the others, namely, 75 each. Finally, when E is 300, both the first and second claimants get half their due, 50 and 100, respectively, with the remaining 150 going to the largest claimant. Thus, what at first appeared to be a capricious subdivision of the estate now seems natural in retrospect. An inspection of Table 2.3 also reveals that the divisions satisfy the consistency requirement.

The arbitration rule undergoes a subtle shift in interpretation as E passes $D/2$. For E below that level, one thinks of the x_i as awards but beyond $D/2$ the focus is on the loss $d_i - x_i$ that is incurred. This finds a resonant echo in the Talmudic script where one finds, in essence, that "less than half is nothing, more than half is all." In effect, getting half the lien is worthless and can be written off so any award is found money. However, getting more than half is frustrating since one begins to hope of achieving full restitution and anything less than that is a deprivation.

When the inheritance rule is applied to congressional apportionment it meets some of the criteria of fair apportionment in the sense that any apportionment acceptable to all states remains acceptable to any subset of states being considered. Hamilton's apportionment scheme, discussed in Section 2.3, violated this criterion, although it was met by the proposals of Jefferson and Webster. An example of this lack of uniformity is illustrated by the case of Oklahoma, which became a state in 1907. Up to that time, the House consisted of 386 seats, apportioned, according to

Hamilton's method, so that New York had 38 members and Maine 3. Oklahoma's entry entitles it to 5 seats, which it received, and the new House size was 391. However, under Hamilton's method, New York would now be forced to give up one seat in favor of Maine for a total of 37 in New York and 4 in Maine, even though the population in each state had not changed.

2.5. Exercises

2.5.1. An algorithm for minimizing (2.3) subject to the constraining relations (2.2) is not hard to implement. From (2.2) we observe that

$$x_j = r_j - x_{j+7} - x_{j-1} < r_j - x_{j-1}$$

in which x_0 is set equal to x_7 and $j = 1, 2, \ldots, 7$. The same relations also show that

$$x_j < r_{j+1}$$

for all j. Now define

$$x_j = \min(r_j - x_{j-1}, r_{j+1}) \qquad (2.15)$$

in which r_8 is set equal to r_1. The last equation in (2.2) shows that

$$x_7 < \min(r_1, r_7) \qquad (2.16)$$

and so it follows that x_j are nonnegative for all j. The algorithm begins by choosing an initial value of x_0, namely, x_7, in relation (2.15), to compute x_1. From (2.16) there are only a finite number of choices. Since x_j takes on the largest possible value for each $j = 1, 2, \ldots, 7$, it suffices to apply (2.15) for each initial choice of x_0 and compute the corresponding sum of the x_j to obtain the largest possible sum. The same choice also minimizes (2.3) because (2.15) always picks the value $r_j - x_{j-1}$ (unless constrained to choose the smaller value r_{j+1}) and this then forces x_{j+7} to be zero. Otherwise x_{j+1} is nonnegative (show this!).

Try this procedure for the case in which $r_1 = 5, r_2 = r_5 = 2, r_3 = 6, r_4 = r_6 = 3$, and $r_7 = 1$. We see that x_0 is either zero or one and so there are two possible solutions to the integer program.

2.5.2. Find the maximum of the weighted sum 2.6 subject to the constraints 2.4 to obtain a solution to the sanitation problem. Recall that $r_1 = 2, r_2 = 3$, $r_3 = 4, r_4 = r_5 = r_6 = 7$, $r_7 = 30$. Show that the solution is the same as maximizing the number of three-day weekends and write down a 30-week rotating schedule for any given worker. Show that $x_9 = 9$.

Note that we seek integer solutions x_j that are nonnegative. Show, for arbitrary choices of the r_i, that the largest number of three-day weekends, namely, the maximum of $x_7 + x_8$, is given by

$$\min(r_1, r_2) + \min(r_5, r_6).$$

2.5.3. By permuting rows in Table 2.2 show that it is possible to arrange for two three-day weekends and five two-day weekends, without requiring six-day workweeks.

2.5.4. Establish the inequalities (2.11) that define Webster's apportionment method. Hint: employ reasoning similar to that used in deriving (2.10) in the text.

2.5.5. Suppose that N customers share a public facility such as a reservoir. The cost of the service must be allocated fairly among the users by a system of fees. This is a topic that is not unrelated to the apportionment problem and it has a number of ramifications (see, for example, the book *Cost Allocation* edited by P. Young, Elsevier North-Holland Press, 1985). We touch on only one aspect here.

Let S be a subset of the N users and $C(S)$ the least cost of servicing all the customers in S most efficiently. If S is empty, then the cost is zero. Moreover, if x_i is the fee charged to the ith customer we impose the breakeven requirement that

$$\sum_{i=1}^{N} x_i = C(N)$$

Because the N customers are cooperating in the venture of maintaining the utility it seems reasonable that no participant or group of participants should be charged more than the cost required to provide the service:

$$\sum_{i \text{ in } S} x_i \leq C(S) \tag{2.17}$$

A related idea is that no participant or subset of participants should be charged less than the incremental cost of including them in the service. Since the incremental cost is $C(N) - C(N - S)$ one requires that

$$\sum_{i \text{ in } S} x_i \geq C(N) - C(N - S) \tag{2.18}$$

Inequality (2.17) expresses the incentive for voluntary cooperation among the participants, whereas (2.18) is a statement of equity since a violation of this inequality by some subset S of users means that S is being subsidized by the remaining $N - S$ customers. In effect, these inequalities express the idea that the individuals have formed a coalition to share both costs and benefits fairly. Show that (2.17) and (2.18) are equivalent.

2.5.6. A single day is broken down into six workshifts for the police department in a certain town to cover manpower requirements, which vary at different times:

Shift	Police required on duty
midnight–2 A.M.	6
2 A.M.–8 A.M.	4
8 A.M.–noon	5
noon–4 P.M.	6
4 P.M.–6 P.M.	8
6 P.M.–midnight	10

Each member of the police force works a consecutive 8-hour period and so only four feasible work clusters exist that satisfy this requirement, as one readily verifies from the table. Let $a_{i,j} = 1$ if the ith shift appears on the jth work period of 8 hours duration, with $a_{i,j} = 0$ otherwise. Define x_j to be the number of police working if the jth work period is chosen and zero otherwise. Formulate and solve the integer programming problem of minimizing the number of police officers needed each day subject to the condition that the ith shift is to appear on some 8-hour period or other at least n_i times, where n_i is the number of police required on duty during the ith shift, as given in the table. If each person works exactly five days a week, what should the total workforce be to cover an entire workweek?

2.5.7. Show that that the arbitration rule (2.14) is uniquely determined when extended to three persons so as to be an internally consistent rule.

2.6. Further Readings

The revised New York City sanitation work schedules were agreed on in a new contract in 1971 and, according to the *New York Times* ("City wants to redeploy sanitationmen for efficiency," February 11, 1971, and "Sanitation union wins $1,710 raise in 27 month pact," November 17, 1971) "city officials described the new pact as a major breakthrough in labor relations in the city because salary increases were linked to specific provisions to increases productivity." The complete story is told in the mechling paper [38]. A review of the mathematics of manpower scheduling is by Bodin [14].

The apportionment problem for the U.S. Congress is captivatingly discussed in the book by Balinski and Young [5]. One of the pitfalls of modeling optimization questions was revealed in Section 2.3 in which we saw that a formal minimum provided little or no guidance in resolving the underlying issue of fairness.

The reapportionment, or political districting, problem is illustrated by the troubled experience of California which had gained 7 new seats in the last census to give it a total delegation of 52 members to Congress in 1992 ("California is torn by political wars over 7 new seats," *New York Times*, March 3, 1991). There is a similar example from New York ("U.S. court voids N.Y. congressional district drawn for Hispanic," *New York Times*, February 27, 1997).

The inheritance problem from the Talmud is thoroughly treated in [3] and [41].

While the City Burns

3.1. Background

What began as a flareup on the kitchen stove quickly spreads to the wood counter and smoke fills the room. An alert homeowner calls the emergency fire department number and within minutes a fire engine company, siren blaring, shrieks to a halt in front of the apartment and the blaze is shortly under control. Scenes like this are enacted daily in cities and towns all over the country and are familiar enough. But sometimes the consequences are more dire and there could be considerable loss of life and property. To reduce, if not entirely eliminate, these losses is the primary mission of a fire department, whose size and operation must be adequate to match the level of service required of it by the public. Insurance codes and municipal regulations stipulate a minimum of coverage, to be sure, but during periods of escalating demand, especially in inner-city ghettos, the number of incidents stretch the available resources. City budget cutbacks and fiscal belt tightening make it less probable that more fire-fighting equipment and personnel can be added to the existing departmental roster, and it forces the chief and his aides to rethink the deployment of the forces currently at their disposal to make them more effective. Although we focus here on fire services, the same predicament faces virtually all public emergency services including ambulance, police, and repairs.

The number of fire companies and their geographic distribution affects the ability to respond to an alarm in a timely manner, but it is difficult to assess by just how much. A doubling of fire companies is likely to reduce response time, for example, but the magnitude of effect is arguable. Nevertheless, response time, namely, the time from when an alarm is called in to the moment that the first fire company arrives at the incident, remains a useful proxy measure of how a redeployment of fire-fighting units can reduce property losses and fatalities. This leads to the problem of how to allocate fire companies to different portions of a city so as to minimize response time, given that the total number of fire-fighting units is fixed. This is considered in Sections 3.3, 3.4, and 3.5.

41

Two kinds of units are typically involved, an engine company consisting of a pumper truck and the men and women assigned to it, that hook up to a fire hydrant to deliver water, and a ladder company that does rescue work by breaking into a burning building. Similar considerations apply to both and we will not distinguish between them.

Alarm rates fluctuate throughout the day, and they also vary considerably from high density areas in the urban core to the more sparsely populated regions at the edge of the city. Moreover during peak alarm periods, often in late afternoon and evening, some fire companies could be busy and consequently not available to respond immediately to a new alarm. The time a unit is busy, from the moment of its initial dispatch until it is again available for reassignment, is also somewhat unpredictable. For these reasons, the deployment problems will be modeled in terms of random variables and the mathematics needed is reviewed in the next section.

Fire companies are placed in firehouses (sometimes more than one to a house) that are located in sectors of the city where fires are expected to occur. Shifting demographics over the years have rendered some of these locations to be less favorably positioned than they were intended to be originally, but, by and large, more companies are located in high-demand areas such as the central business district and less, say, in residential areas. However, a residential area may contain high-risk fire hazards, such as schools, and its residents are penalized for low incidence rates by more dispersed fire companies and, therefore, greater response times. This inequity in coverage can be adjusted by shifting some of the units in high alarm areas to the residential zones but this creates an imbalance in the workload of the fire-fighting units since each responds to more calls than those in the more sparsely populated part of town. We therefore see that there are several, possibly conflicting, "fairness" criteria that are reminiscent of the multiple objectives encountered in the previous chapters in different guises. The trade-off between these goals is discussed again in Section 3.5 together with other deployment issues. The analysis of these problems is based on work done by the Rand Institute in New York City about three decades ago (see the references in Section 3.7).

This chapter draws on the notion of conditional probability, which is briefly reviewed in the book's appendix.

3.2. Poisson Processes

A sequence of events occurs randomly in time and we count the number $N(t)$ that have taken place up to time t. With an eye toward the applications later in this chapter, we think of $N(t)$ as the number of fire alarms that arrive at some central dispatcher (either by phone or through fire-alarm call boxes) by time t. $N(t)$ is a nonnegative and integer-valued random variable that satisfies the relation $N(s) \leq N(t)$ for $s < t$, with $N(0) = 0$.

We assume that the number of calls that arrive in disjoint time intervals are statistically independent. This assumption is called *independent increments* and may not be quite true for fire alarms since a call that is not answered immediately could allow a minor flareup to escalate into a serious fire that would trigger a flurry of other calls over a span of time. Nevertheless it seems to be a reasonable hypothesis most of the time.

We also assume that the probability distribution of the number of arrivals that take place in any time interval depends only on the length of that interval and not on when it begins. In other words, the number of calls in the interval $(t_1 + s, t_2 + s)$, namely, $N(t_2 + s) - N(t_1 + s)$, is distributed in the same way as the number of calls that take place in (t_1, t_2), namely, $N(t_2) - N(t_1)$, for all $t_1 < t_2$ and $s > 0$. This condition, called *stationary increments*, is violated for fire alarms since the frequency of fires varies with the time of day. However, one can reconcile this hypothesis with real data by restricting our observations to peak alarm periods when the arrival rate of calls is fairly constant.

Let us now define a *Poisson process* (after the French mathematician S. Poisson) to be a counting process with stationary and independent increments for which the probability distribution of the random variable $N(t)$ is

$$\text{prob}(N(t) = k) = \frac{(\lambda t)^k e^{-\lambda t}}{k!} \tag{3.1}$$

for $k = 0, 1, \ldots$. The constant λ is called the *rate* of the process for reasons to become clear in a moment. Note that by the stationary increments requirement we have

$$\text{prob}(N(t + s) - N(s) = k) = \text{prob}(N(t) = k)$$

Let $o(t)$ denote terms in t of second or higher order. These are negligible terms when t is small enough. From (3.1) we obtain

$$\text{prob}(N(t) = 1) = \lambda t e^{-\lambda t} = \lambda t \left(1 - \lambda t + \frac{(\lambda t)^2}{2} - \cdots \right) = \lambda t + o(t)$$

and, in a similar manner,

$$\text{prob}(N(t) > 2) = 1 - \text{prob}(N(t) = 0, 1) = o(t)$$

These relations show that if t is small enough, the probability of more than one arrival in an interval of duration t is negligible and the probability of a single arrival is roughly equal to λt.

The expected value of $N(t)$ is easily computed to be λt (Exercise 3.6.1), which enables us to interpret λ as the average number of events per unit time. This, and a number of other facts about the Poisson distribution, can be found in any number of books on stochastic processes (see, for example, the book by Ross [46]) or are left as exercises.

When m Poisson processes are taking place simultaneously and independently it is not unreasonable that the sum is also Poisson.

Lemma 3.1   *Let $N_i(t)$ be independent Poisson random variables at rates λ_i, for $i = 1, 2, \ldots, m$. Then the sum $N(t) = N_1(t) + N_2(t) + \cdots + N_m(t)$ is also Poisson at rate $\lambda = \lambda_1 + \lambda_2 + \cdots + \lambda_m$.*

Proof: Begin with the case $m = 2$. Since the event $N(t) = k$ occurs in $k + 1$ disjoint ways, namely, $N_1(t) = i$, and $N_2(t) = k - i$, for $i = 0, 1, \ldots, k$, we can sum over the $k + 1$ events to obtain

$$\text{prob}(N(t) = k) = \sum_{i=0}^{k} \text{prob}(N_1(t) = i, N_2(t) = k - i)$$

(see the appendix). Moreover, the separate counting processes are statistically independent, and so the last sum becomes $\sum_{i=0}^{k}(\lambda_1 t)^i e^{-\lambda_1 t}(\lambda_2 t)^{k-i} e^{-\lambda_2 t}/i!(k-i)! = t^k e^{-(\lambda_1+\lambda_2)t}/k! \sum_{i=0}^{k} k!(\lambda_1)^i(\lambda_2)^{k-i}/i!(k-i)! = t^k(\lambda_1+\lambda_2)^k e^{-(\lambda_1+\lambda_2)t}/k!$, using the binomial theorem. Now proceed by induction. If the sum of the first $m - 1$ processes is already Poisson then all we need do is to add the last one to this sum, which is again the case of two Poisson processes considered above.

A continous random variable T is said to be *exponentially distributed* at *rate* μ if

$$\text{prob}(T \le t) = \begin{cases} 1 - e^{-\mu t}, & t \ge 0 \\ 0, & t < 0 \end{cases} \tag{3.2}$$

The expected value of T is readily verified to be $1/\mu$ (Exercise 3.6.1).

Let $T_1, T_2, \ldots$ denote the length of time between successive events of a Poisson process. A little thought shows that

$$\sum_{i=1}^{k} T_i \le t \quad \text{if and only if } N(t) \ge k \tag{3.3}$$

In the special case of $k = 1$ we therefore see that

$$\text{prob}(T_1 < t) = 1 - e^{-\lambda t}$$

and so T_1 is exponentially distributed.

It is true in general, moreover, that the T_i are independent and identically distributed exponential random variables. This result is not proven here since it is used only incidentally below, although it is not really difficult to do (see, for example, Chapter 5 of the book by Ross [46]). Because of this fact, there is no need to use subscripts to indicate which gap is being considered, and T will denote the interval between any two arrivals.

Suppose that one begins to observe a Poisson process at some random time $s > 0$ and then wait until the first call arrives. The time gap between two successive arrivals is interrupted, so to speak, by the sudden appearance of an observer. It is a remarkable fact that the duration of time until the next event as seen by the observer has the same exponential distribution as the gap length of the uninterrupted interval. This is called the *memoryless property* since it implies that the past history of the process has no effect on its future. In more mathematical terms, this is expressed by saying that the conditional probability of a gap length T greater than $t + s$, given that no call took place up to time s, is the same as the unconditional probability of a gap exceeding t:

$$\text{prob}(T > t + s \mid T > s) = \text{prob}(T > t) \quad \text{for all } s, t > 0 \qquad (3.4)$$

Using the definition of conditional probability (see the appendix for a review), (3.3) is equivalent to

$$\frac{\text{prob}(T > t + s, T > s)}{\text{prob}(T > s)} = \text{prob}(T > t)$$

or, to put it another way,

$$\text{prob}(T > t + s) = \text{prob}(T > t)\,\text{prob}(T > s)$$

The last identity is certainly satisfied when T is exponentially distributed.

Suppose there are two concurrent and independent Poisson processes. A patient observer will see the next arrival from either the first or the second process. The probability that the next occurrence is actually from process i, $1 \le i \le 2$, is λ_i / λ, where λ is the sum $\lambda_1 + \lambda_2$. We prove this in terms of the interarrival times T and T' of two simultaneous Poisson processes: $\qquad\qquad\qquad\qquad\qquad \square$

Lemma 3.2 *Let T and T' be independent and exponentially distributed random variables at rates μ_1 and μ_2. These define the interarrival times from two Poisson processes at rates μ_1 and μ_2. The probability that the first arrival occurs from the process having rate i is μ_i / μ where μ is the sum of μ_1 and μ_2.*

Proof: Let $i = 1$. A similar argument then applies to $i = 2$. We need to compute $\text{prob}(T < T')$. The density function of the variable T' is $\mu_2 e^{-\mu_2 t}$ and so by conditioning on the variable T' (see the appendix), we obtain

$$\text{prob}(T < T') = \int_0^\infty \text{prob}(T < T' \mid T' = s)\mu_2 e^{-\mu_2 s}\,ds$$

However, since T and T' are independent, the last integral becomes

$$\int_0^\infty \text{prob}(T < s)\mu_2 e^{-\mu_2 s}\, ds = \int_0^\infty (1 - e^{-\mu_1 s})\mu_2 e^{-\mu_2 s}\, ds = \frac{\mu_1}{\mu}$$

The time from when an alarm is received by a dispatcher until a fire company arrives at the incident and completes its fire-fighting operations is called the *service time*. We assume that the successive service times of a particular fire company define a Poisson process in which the kth "arrival" occurs when the kth service is complete, ignoring idle periods during which the company is not working. From now on we use the more appropriate term "departure" since our concern is with completion of service, and the preceding discussion shows that consecutive departure times are independent exponential random variables. Service times of a fire company are in fact not exponentially distributed in general, but the results obtained by making this assumption give results that are reasonably close to those obtained in practice, as will be seen later.

The fire companies operate more or less independently, and if m of them are busy at incidents this constitutes m independent and identically distributed Poisson processes at rates μ (that is, average service times $1/\mu$). From Lemma 3.1 their combined rate is $m\mu$ and so the average time required for some company or other to become the first to complete its service is then $1/m\mu$. If a new alarm arrives from a Poisson process at rate λ while the m units are busy, then, in view of the *memoryless property* of the exponential, the service time remaining from the receipt of the call is again exponential at a mean rate of $1/m\mu$.

Now suppose that a municipality has a total of N fire companies and that they are all are busy. The arrival of a new alarm at rate λ is unconnected with the departure of a call presently in service, and so we have two independent and concurrent Poisson processes at rates λ and $N\mu$. Lemma 3.2 now shows that the probability that the new call must wait for a busy unit to become available (namely, that an arrival occurs before a departure) is $\lambda/(\lambda + N\mu)$.

Imagine that the peak alarm period has been going on for some time so that incoming calls and service completions have reached a sort of equilibrium in which the average number of alarms that arrive equals the average number of departures. In this case, it is evident that if the average number of busy units is M, then $M\mu$ is the average departure rate from the system. Since this equals the mean arrival rate λ, we obtain the relation

$$M = \frac{\lambda}{\mu} \tag{3.5}$$

Formula (3.5) is well known in queueing theory, which is the mathematical study of waiting lines, and can be established rigorously under fairly general hypothesis. We will have more to say about relation (3.5) in Section 3.4.

A spatial Poisson process is defined in a way similar to a temporal process. Suppose events take place in the plane at random and that if S is any subset of the plane, then $N(S)$ counts the number of events that occurs within S. If S and S'

are disjoint subsets then we require that $N(S)$ and $N(S')$ be independent random variables ("independent increments") and that $N(S)$ depend only on the area of S and not on its position or shape ("stationary increments"). When E is the empty set then $N(E) = 0$.

The spatial counting process is called Poisson if

$$\text{prob}(N(S) = k) = \frac{(\gamma A(S))^k e^{-\gamma A(S)}}{k!}$$

where $k = 0, 1, \ldots$ and $A(S)$ is the area of S. The rate constant γ is easily shown to be the average number of events per unit area, using an argument analogous to that employed in the temporal case.

It is useful to know how to estimate the rate constant λ in a Poisson process. Break a time interval t into n small pieces of size h, so that $t = nh$, and count how many arrivals actually occur during t. Call this m. By independent increments what happens in a given interval is independent of whether or not an arrival occurs in any other interval. The probability of an arrival in an interval of length h is one minus the probability of no arrival which, by the Poisson distribution, is roughly λh, as we saw earlier, with the approximation getting better as h tends to zero for fixed t or, equivalently, as t gets larger when h is fixed. Moreover the probability of more than one arrival during h is roughly zero, for the same reason. Therefore we have a sequence of independent trials with two outcomes, arrival or nonarrival (the well-known Bernoulli trials), and so the average number of arrivals during t is $n\lambda h$. It follows that

$$m = n\lambda h = t \qquad \text{or} \qquad \lambda = \frac{m}{t} \quad \text{for large enough } t$$

For spatial processes the same result holds in the form $\gamma = m/A(S)$ for planar regions S of sufficiently large area. □

3.3. The Inverse Square Root Law

A large region S of area $A(S)$ has N firehouses scattered at random according to a spatial Poisson distribution at rate γ. The rate constant is the average number of firehouses per unit area and, according to our previous discussion, is estimated to be $N/A(S)$. Alarms arrive as a Poisson process at rate λ and any fire company that is dispatched remains busy for a length of time that is exponentially distributed with a mean duration of $1/\mu$. We assume one fire company per house, a requirement that is modified later, and that calls arrive during peak alarm periods during which the mean λ can be considered to be constant. The more quiescent periods of the day, if there are any, are less interesting since the demand for service can be satisfied more readily.

An incident occurs somewhere at random, uniformly distributed within S, and the closest available unit is dispatched. Implicit here is that alarms are just as likely to happen in one part of the region as another. This implies that S is fairly homogeneous in terms of hazards, as would be the case in a number of residential areas. For the moment it is also assumed that all N units are available, a restriction that will be relaxed later to account for the fact that some fire companies may be busy at other alarms.

Let us think of S as a portion of a metropolis in which the street network is a grid of criss-crossing intersections. This means that travel to an incident is not along the shortest path between two points, or "as the crow flies," but along a less direct route. We want to make this idea more precise by introducing a measure of distance in the plane that is called the *right angle metric*. This defines the distance between the origin and a point having coordinates (x, y) to be $|x| + |y|$. Travel in this metric is along a horizontal distance $|x|$ followed by a vertical distance $|y|$, which is different from the conventional Euclidean metric distance defined by $\sqrt{x^2 + y^2}$. In an actual street pattern, the distance traveled would be somewhere between these extremes but at the risk of being too conservative we adopt the right angle metric. Later computations will reveal that there is not a significant difference between the two (Exercise 3.6.3).

The locus of points in the x, y plane that are within a distance r of the origin is of course a circle of radius r, when using the Euclidean metric, and the area is πr^2. But, as a little thought will show, the corresponding locus of points that are within a distance r using the right angle metric is now a square tilted by 90 degrees with sides of length $\sqrt{2}r$ and area $2r^2$ (Figure 3.1).

If the travel speed of a fire truck is roughly constant, then response time is proportional to travel distance and so from now on we work with distance traveled rather than elapsed time.

At this juncture, we are able to compute the probability distribution of the travel distance D_1 between an incident and the closest responding fire company (see Figure 3.2 for an illustration of a typical situation). Choose the coordinate system so that the incident is located at the origin. Since firehouses are Poissonly distributed in the region, the probability of there being no fire company within a distance r of the incident is simply $e^{-2\gamma r^2}$, using the fact that area in the right angle metric is $2r^2$. Therefore the probability that the closest unit is within a distance r is given by

$$F(r) = \text{prob}(D_1 \le r) = 1 - \text{prob}(D_1 > r) = 1 - e^{-2\gamma r^2} \qquad (3.6)$$

The density function of the random variable D_1 is obtained from (3.6) by differentiating $F(r)$ and is given by $F'(r) = 4\gamma r e^{-2\gamma r^2}$. The expected value of D_1 is now computed by the integral

$$E(D_1) = \int_0^\infty r F'(r)\, dr = 4\gamma \int_0^\infty r^2 e^{-2\gamma r^2}\, dr \qquad (3.7)$$

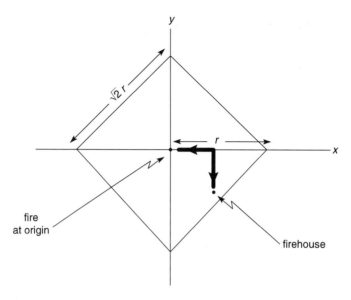

Figure 3.1. The right-angle distance metric showing a fire incident at the center, a fire-house, and a right angle travel path from the firehouse to the incident (as dark line).

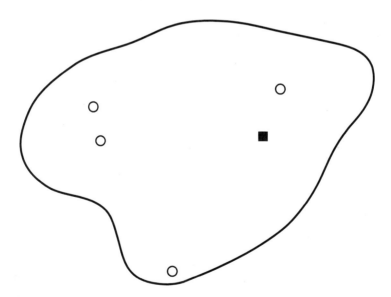

Figure 3.2. A hypothetical region with Poisson distributed fire companies (open circles) and an incident (dark square).

which is evaluated by rewriting (3.7):

$$E(D_1) = -2\gamma \frac{d}{d\gamma} \int_0^\infty e^{-2\gamma r^2}\, dr$$

Now make the substitution $x = \sqrt{(2\gamma)}r$ to obtain

$$\int_0^\infty e^{-2\gamma r^2}\, dr = \frac{1}{\sqrt{2\gamma}} \int_0^\infty e^{-x^2}\, dx$$

and from a table of integrals we find that the last integral has the value $\sqrt{\pi}/2$. Putting all this together gives

$$E(D_1) = -\gamma \left(\frac{\pi}{2}\right)^{1/2} \frac{d}{d\gamma}\left(\frac{1}{\gamma^{1/2}}\right) = 0.627\gamma^{-1/2} \tag{3.8}$$

We therefore see that the expected travel distance is inversely proportional to the square root of the density of available units. This relation persists under a variety of assumptions about the distribution of response vehicles in the region and the metric one employs. It even continues to be valid if the kth closest fire company is dispatched instead of the nearest one. We can see this easily enough in the case of D_2, the distance to the second closest unit. The probability that the second closest is within a distance r of an incident, using the right angle metric as before, is one minus the probability that either none or exactly one unit is within r. From the Poisson assumption about the distribution of firehouses we obtain

$$G(r) = \mathrm{prob}(D_2 \le r) = 1 - e^{-2\gamma r^2} - 2\gamma r^2 e^{-2\gamma r^2}$$

with a density function given by $G'(r) = 8\gamma^2 r^3 e^{-2\gamma r^2}$.

A computation like before (Exercise 3.6.4) yields

$$E(D_2) = \int r G'(r)\, dr = 0.941\gamma^{-1/2} \tag{3.9}$$

Therefore, except for a slightly larger constant of proportionality, the expected value is the same as (3.8).

Because $\gamma = N/A(S)$, we can rewrite expressions (3.8) and (3.9) as $c_1/N^{1/2}$ and $c_2/N^{1/2}$ for suitable constants, to make the dependence on N more explicit. These mean values assume that all N units are available to respond, which is unrealistic since some fire companies could be busy on other calls. Let m be the number of busy units. In Section 3.2 we showed that the mean of the random variable m is λ/μ

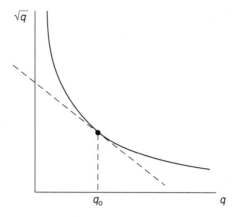

Figure 3.3. The convex function $1/\sqrt{q}$, with tangent line at some point q_o.

and so the expected value of the variable $q = N-m$ is $E(q) = N-\lambda/\mu$. Conditioning on q, the number of units actually available, the expectation of D_1 is given by

$$E(D_1 \mid q) = c_1/q^{1/2}$$

It is a standard result in probability theory (see the appendix) that the unconditional mean of D_1 is the average of $E(D_1 \mid q)$ with respect to q, namely,

$$E(D_1) = E(E(D_1 \mid q))$$

Now $1/q^{1/2}$ is, for $q > 0$, a convex function, meaning that the tangent line to the curve defined by the graph of the function lies on or below the curve itself (Figure 3.3). Using simple calculus one shows that, because of the convexity,

$$E(D_1) = E(E(D_1 \mid q)) \geq \frac{c_1}{E(q)^{1/2}} \qquad (3.10)$$

(Exercise 3.6.6). If one is willing to hazard a possibly low estimate for the mean of D_1, the inequality in (3.10) can be replaced by an equality to give an expression for $E(D_1)$ in terms of the average number of available units:

$$E(D_1) = \frac{c_1}{(N - \lambda/\mu)^{1/2}} \qquad (3.11)$$

The same relation holds for D_2 except that the constant is replaced by c_2.

Empirical verification of (3.11) has been obtained by plotting actual response time data against the number of available units (see the paper by Kolesar and Blum [34]), and it is remarkable that the relation persists in spite of the several tenuous assumptions that were made in its derivation, some acceptable, others perhaps less so, such as constant vehicle speed, uniform alarms in space, exponential service times, and a rectangular street grid. The robustness of (3.11) under a variety of conditions compels us to call it the *inverse square root law*, and it provides a simple link between available fire-fighting resources and the ability to respond to an alarm.

3.4. How Busy Are the Fire Companies?

Suppose that a district has S fire companies situated within its boundary. When there is a serious multiple-alarm blaze it is conceivable that all these fire-fighting units would be simultaneously occupied in an attempt to contain the conflagration. During this time, any additional calls for service would have to be handled by more distant units from outside the district, if any are available, and the response time would be longer than usual.

Events that actually strip a district of its fire-fighting resources are not that unusual in a large city. An explosion in a factory that rages out of control is an example, and the many fires deliberately set during an urban riot is another. Under circumstances like these, it is of interest to know what the probability would be that all S units are busy and, hence, unavailable to other alarms.

To set up the framework for our analysis of this problem we consider first the simpler case in which exactly one unit is dispatched immediately to each call. In this setting, the number of alarms being serviced is identical to the number of busy fire companies, which virtually excludes from consideration all serious alarms that require multiple servers. We then consider the more realistic situation in which alarms pass through *stages*. A stage is a period of time during which a fixed number of fire engines are busy at a given incident. When a small fire occurs, for example, three engines might be dispatched and then two released as soon as the first one arrives at the scene and determines that the fire is not serious. Such an incident has two stages. The first stage represents the time until the first engine arrives after receipt of an alarm. The second stage represents the remainder of the incident and has one busy engine. A large fire can generate many stages. For example, suppose that three are initially dispatched. When the first unit arrives at the scene it determines that the fire is serious and calls for two more units. This initiates a second stage. When the fire is brought under control four out of the five busy units can be released, allowing a single unit to complete the mop-up operations. The moment of release is the beginning of a third and final stage. It is apparent, then, that the number of busy fire companies at any time is not necessarily the same as the number of fires in progress.

Let us begin with single-stage fires that engage only one fire-fighting unit. When the fire is serious, what would actually happen is the first unit arrives and then calls for an additional $k - 1$ units to be sent to the same location but, for the sake of simplicity, we temporarily set k to unity and collapse the separate stages into a single one.

The ensuing discussion will use the material in Section 3.2 as a point of departure. We assume, as usual, that alarms arrive as a Poisson process at mean rate λ and

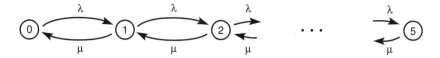

Figure 3.4. Transition diagram of flow between states that represent the number of busy fire companies.

the time each fire company is busy on a call is exponential at rate μ. A simplifying assumption is also made that the system of alarms and responses has reached a steady state. This simply paraphrases the *equilibrium condition* expressed in Section 3.2 that during peak alarm periods the average number of incoming alarms is balanced by the average number of service departures. We use this to imply that the probability of finding exactly k busy units is independent of time. In effect we have a Markov chain in which the long-term probability of finding the system in a particular state no longer varies with the time at which transitions take place. Whether it is in fact possible to disregard fluctuations in these probabilities depends on if the peak alarm period lasts long enough to ignore initial transient effects and if there are no unexpected disruptions to service in the interim.

The number of busy fire companies is $0, 1, \ldots, S$ and we agree to call these the *states* of the fire-reponse system. When state S is attained, no additional alarms can be serviced within the district and, as far as we are concerned, these may be considered as "lost" calls. Transitions between states are computed by using the equilibrium hypothesis. Consider the *transition diagram* shown in Figure 3.4, which expresses the "flow" between states, depending on whether a new alarm has been received at rate λ, or a busy unit has completed service at rate μ, and is then again available to respond to other calls.

In an equilibrium setting, the total flow into and out of a given state must be conserved. For example, the rate at which state 1 changes into either state 0 or 2, namely, $\lambda + \mu$, must equal the rate λ into 1 from 0 plus the rate 2μ from 2 into 1. If p_k denotes the probability of being in state k, then the appropriate balance relation for state 1 is that $(\lambda + \mu)p_1 = \lambda p_0 + 2\mu p_2$, because this conditions the average transitions between states on the probability of actually being in those states. The same argument applies to all other states, and one obtains

$$\lambda p_0 = \mu p_1$$
$$(\lambda + k\mu)p_k = \lambda p_{k-1} + k\mu p_{p+1}, \quad k = 1, \ldots, S \tag{3.12}$$

The system of equations (3.12) is easily solved recursively by writing p_1 in terms of p_0 and then p_2 in terms of p_1, and so forth. From this we get $p_k = \rho^k p_0/k!$, where $\rho = \frac{\lambda}{\mu}$. Because the system must be in exactly one of the given states the sum of the p_k must sum to unity and from this we find that

$$p_k = \frac{\rho^k \left(\sum \rho^n/n! \right)^{-1}}{k!} \tag{3.13}$$

Table 3.1. Values of p_k for a
district with $S = 15$ units.

k	p_k
0	.0003
5	.0916
8	.1396
10	.0993
12	.0481
15	.0090

When k is large enough the sum can be approximated by an exponential and we obtain the Poisson approximation

$$p_k = \frac{\rho^k}{k!} \exp(-\rho) \tag{3.14}$$

In particular, the probability of all S units being simultaneously busy is obtained by letting $k = S$. Because the distribution of busy units is Poisson, it follows that the mean number of busy units is just ρ, a rediscovery of relation (3.5). Table 3.1 computes p_k from 3.14 for a district having $S = 15$ units. The mean arrival rate is taken to be eight calls per hour and service time averages 60 minutes, and so $\rho = 8$.

The probability peaks at $k = 8$, which is also the average number of busy units.

Now consider the somewhat more realistic situation in which there are two stages in which two units are sent initially to every alarm followed by the release of one unit when the fire appears to be still smoldering but under control. The service times in each stage (that is, the duration of the alarms in progress in each stage) are exponential at rates μ_1 and μ_2 and these are deemed to be independent of each other. Stage 1 occurs when an alarm arrives as a Poisson distribution at rate λ and stage 2 begins when one of the two busy units terminates its service. The output from stage 1 is the input to stage 2, and to apply the foregoing analysis we must be sure that the input to the second stage is itself Poisson at the same rate so that what we have in effect are two independent Poisson processes in tandem. This is the case, in fact, as established in Exercise 3.6.15.

There is a trade-off between between sending two units initially and finding that only one is needed, or initially sending one to a fire that may in fact be serious. In the first instance, one of the units is temporarily unavailable to respond elsewhere, whereas in the other situation there is a delay before a much-needed second unit finally arrives. The delay in the availability of fire-fighting units increases the risk of loss of life and property.

To analyze the two-stage system we define states by the doublet (n, m) for nonnegative integers n, m in which n is the number of alarms in progress in stage 1 and m is the number of alarms in progress in stage 2. The phrase "in progress"

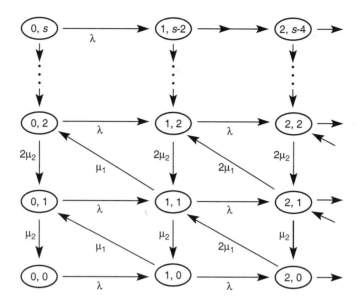

Figure 3.5. Transition diagram of flow between states that represent the number of alarms in progress in each of two stages.

means that the alarms are being serviced by one or two fire units, as the case may be. The transition diagram for the flow between states is shown in Figure 3.5, and Figure 3.6 presents a "closeup" view of the flow into and out of a generic state (n, m).

Let $p(n, m)$ be the equilibrium probability of finding the fire-response system in state (n, m). Then, just as we did before, the following balance relations for the flow in and out of any state can be read from Figure 3.6:

$$(\lambda + n\mu + m\mu)p(n, m) = \lambda p(n - 1, m) + (m + 1)\mu_2 p(n, m + 1)$$
$$+ (n + 1)\mu_1 p(n + 1, m - 1)$$

for $n, m \geq 1$. Similar relations hold when either n or m is zero.

Because there are two units operative in stage 1 and one in stage 2 there will be k busy units altogether whenever n and m satisfy the relation $2n + m = k$. Therefore the equilibrium probability p_k of finding exactly k busy fire companies is given by

$$p_k = \sum\sum p(n, m) \tag{3.15}$$

where the double sum is over all n, m for which $2n + m = k$. The two stages represent independent fire-fighting events with Poisson arrivals and exponential service and so, for all S large enough, the probability of n alarms in progress in stage 1 and m in stage 2 can be approximated by Poisson distributions, by the same

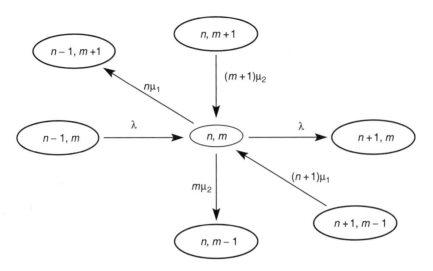

Figure 3.6. A close-up view of the diagram in the preceding figure showing the flow into and out of a generic state.

reasoning used earlier leading to (3.15). Because of independence $p(n, m)$ is the product of the separate distributions and therefore

$$p(n, m) = \frac{\rho_1^n \rho_2^m e^{-r}}{n! m!} \tag{3.16}$$

where $r = \rho_1 + \rho_2, \rho_1 = \lambda/\mu_1$, and $\rho_2 = \lambda/\mu_2$. An easy computation shows that (3.16) satisfies the balance relations shown. The value of p_k is now, from (3.15) and (3.16),

$$p_k = e^{-r} \sum \sum \frac{\rho_1^n \rho_2^m}{n! m!}$$

where, once again, the double sum extends over all n, m such that $2n + m = k$. In particular when $k = S$, we get the probability that all available fire companies are simultaneously busy. Table 3.2 gives the values of p_k when $S = 15$, as before, but with an assumption of 45 minutes service time in stage 1 and 15 minutes in stage 2 for a total of 60 minutes. Thus $\rho_1 = 6$ and $\rho_2 = 2$ and $r = 8$.

The probabilities peak at $k = 13$ at a higher value than that obtained for the single-stage fire treated earlier. Also the likelihood that all 15 fire companies are simultaneously busy is now .1508 which is a roughly 17-fold increase over the same probability (see Table 3.2) when there are no serious fires requiring multiple servers.

Let us enlarge on this observation. In the multiple dispatch strategy two units are sent so that the alarm rate in the equilibrium relation (3.5) is now effectively

Table 3.2. Values of p_k for a district with $S = 15$ units.

k	p_k
0	.0003
5	.0251
8	.0792
10	.1206
12	.1494
13	.1560
15	.1508

doubled compared to the situation in which a single unit responds to each alarm. If the second stage is discounted, the average number of busy units is also doubled. This is another way of stating that the likelihood of stripping the region of all its fire-fighting resources increases when there are multiple stages in which several units can be busy at once.

3.5. Optimal Deployment of Fire Companies

A city has been partitioned into k districts and within each of these fire alarms occur at random in a spatially uniform manner. These homogeneous sectors represent the different alarm histories that one can expect in diverse parts of the overall region, such as high- and low-density residential and business zones, where hazard rates vary with locale. Some sectors may also be determined by geography as, for example, when a river forms a natural barrier that separates one zone from another.

Calls in district i are assumed to come from a Poisson distribution at rate λ_i, $i = 1, 2, \ldots, k$. The probability that the next citywide alarm actually originates from district i is, from Lemma 3.2, λ_i / λ, where λ is the sum of λ_1 through λ_k. For each i, let N_i be the number of fire companies assigned to the ith district and let μ_i be the average service time of companies in that district, which depends on travel conditions and the severity of fires. Service times are exponentially distributed.

Conditioning on the k disjoint events that an alarm arrives from sector i, the regionwide unconditional response distance D to the closest responding unit (our surrogate for response time) has an expected value that can be computed from formula (3.11):

$$E(D) = \sum_{i=1}^{k} \frac{c_i \lambda_i}{\lambda (N_i - \lambda_i / \mu_i)^{1/2}} = \sum_{i=1}^{k} g_i(N_i) \tag{3.17}$$

where c_i equals .627 times the square root of the area of sector i. The fire department would like $E(D)$ to be as small as possible and this leads to the optimization problem of minimizing this sum subject to the conditions that the N_i add up to N, the total available resources in the city, and that N_i are integers, at least one greater than the smallest integer in λ_i/μ_i. Once again, as in Chapter 2, this is an integer program and it admits a simple solution. One begins by assigning the smallest possible number of units, $1 + \lambda_i/\mu_i$, to the ith term in (3.17). If this bare-bones deployment adds up to N, we are done. Otherwise allocate one more unit to that term for which $g(N_i) - g(N_i + 1)$ is largest, because this choice most decreases the sum. Continue in this fashion until all N available units have been assigned.

The resulting allocation is designed to optimize the citywide efficiency of deployment but it may suffer in other respects. For example, a high-demand area with many fire alarms would receive somewhat more companies than would an adjacent sector that has fewer alarms, such as a low-density residential part of town. However, when a fire does occur in the residential area the response time is likely to be greater since the responding units are more dispersed. The tax-paying residents are understandably resentful of this inequity in coverage because they are being penalized for having less fires. Suppose then, in response to their outcry, that some additional units are permanently shifted to the residential zone. This leaves the high-density area more vulnerable than before since travel times would tend to increase and, what is perhaps of equal significance, the workload of the fire companies in the high-demand area would exceed that of the companies in the adjoining region in the sense that each is busier a greater fraction of the time. The firefighters' union would protest.

There is an evident need to reconcile the multiple and often conflicting interests of the city administration, the employees union, and community groups. Although the citywide allocation based on the inverse square root law may fail to adequately meet the criteria of fairness imposed by citizens and firefighters (equity of coverage and equity of workload), it provides a compromise by placing more units in the high-demand areas while ensuring an acceptable level of coverage, and it has been used effectively by several municipalities as a rule of thumb for resource allocation. In New York City, for instance, during one of its recurrent fiscal crises, the fire department budget was cut and some fire companies had to be disbanded. Other units were then relocated to fill the resulting gaps in coverage by employing the inverse square root law to minimize the degradation in service caused by the cuts.

These changes were initially resisted by the firefighters as well as by community groups that felt threatened by the moves, but ultimately the cost-saving measures were implemented (for further comments on this, see Section 3.7).

One way to achieve equity of coverage is to temporarily reposition fire companies to other houses during periods of heavy demand. This relieves busy units and tends to reduce workload imbalances as well. The repositioning problem can be formulated mathematically by first partitioning a region into subzones called "response neighborhoods," each of which is served by the two closest fire companies. Some firehouses may, of course, belong to more than one response neighborhood, assuming one company per firehouse, and so the partition consists of overlapping subzones.

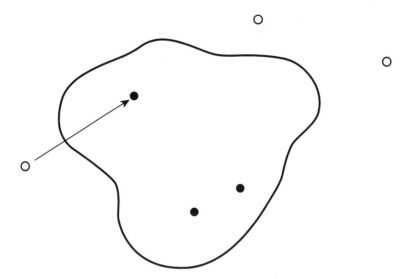

Figure 3.7. Illustration of repositioning with one unprotected response neighborhood and several adjacent firehouses. The dark circles are empty firehouses and the open circles designate firehouses with an available unit. The arrow shows a deployable company being moved, temporarily, to the empty firehouse of a busy company.

A response neighborhood is uncovered if both of its closest units are busy on other calls, in which case some other available unit is temporarily assigned to one of the empty firehouses (Figure 3.7). The problem now is to decide which of the empty houses to fill so as to minimize the number of relocated companies (Exercise 3.6.8). The issue here is not which of the available units to reposition but, rather, how many, and is reminiscent of the integer programming problems considered in the previous chapter.

Part of the appeal of the inverse square root law is that although it is deceptively simple to state and to use, it is a surprisingly effective tool for planning the long-term deployment of fire-fighting resources. However, there are other short-term deployment issues concerning day-to-day operations that are also of considerable interest, and we touch on them briefly next.

Although we have assumed that one always dispatches the nearest unit to an alarm, it is not obvious that sending a more distant unit could sometimes be more effective in the long run. Usually the optimal response is to deploy the closest unit (or units). But during a busy period in a high-demand area this dispatch policy could strip the immediate area of all its fire-fighting resources and a future alarm might well be answered by a unit that is considerably more distant. Therefore, if the closest unit is not dispatched initially to an alarm, the delay in response experienced by the incident could be compensated for by the ability of that unit to be available to future alarm. There is a trade-off here between the short-term advantage in responding quickly to a current alarm and the long-term gain in

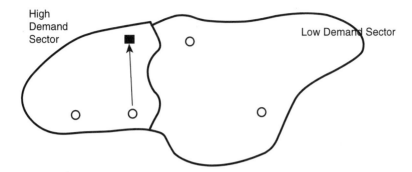

Figure 3.8. Adjacent high- and low-demand areas with the boundary between them drawn so that the first responding unit is not necessarily the closest unit. The open circles are available firehouses and the dark square is an incident. The arrow shows a deployable company responding to the incident.

having units available to send to future incidents. The problem is to decide on how to place the boundary of the response sector of a first responding unit so that an exceptionally busy company has a smaller response zone than would a company in a lower alarm area. With appropriately redefined boundaries any workload imbalance is reduced, and there would be an occasional exception to the rule of sending the closest unit (see Figure 3.8 for an illustration of this idea). This is in contrast to the repositioning problem considered earlier in which boundaries were drawn to specifically include the closest (and second closest) units.

These tactical questions of how many and which units to dispatch can be formulated mathematically but we do not do so here (see, however, the references in section 3.7).

3.6. Exercises

3.6.1. The expected value of a Poisson random variable $N(t)$ at rate λ is λt and if T is exponentially distributed at rate μ, then the expected value of T is $1/\mu$. Establish these relations.

3.6.2. Let T_i be the successive gaps in the arrivals of a Poisson process at rate λ. Using relation (3.4) show that

$$\text{prob}\left(\sum_{i=1}^{k} T_i > t\right) = \sum_{j=0}^{k-1} \frac{(\lambda t)^j e^{-\lambda t}}{j!}$$

with an expected value of k/λ.

3.6.3. Formula (3.8) was derived assuming a right angle distance metric. Using very similar arguments, rederive the expected value of D_1 in terms of the Euclidean metric, in which the distance between two points in the plane is the square root of the sum of the squares of the distances along each coordinate direction. Show that $E(D_1) = .5\gamma^{-1/2}$.

3.6.4. Complete the derivation of formula (3.9) for $E(D_2)$. Extend this result to obtain the expected distance to the kth nearest vehicle.

3.6.5. Compute the variance of D_1 in the right angle metric. Hint: $\text{Var}(D_1) = E(D_1^2) - E(D_1)^2$ and $\text{prob}(D_1^2 \leq r) = \text{prob}(D_1 \leq \sqrt{r})$.

3.6.6. In Section 3.3 we saw that $1/q^{1/2}$ is a convex function (Figure 3.3). Denote this function by $h(q)$. To establish the inequality (3.10) in the text, recall that the convexity of h means that the tangent to the curve $(q, h(q))$ in the plane always lies on or below the curve itself. If $q_o > 0$ is some given point, the equation of the tangent line at q_o is $t(q) = h(q_o) + h'(q_o)(q - q_o)$, where $h'(q_o)$ denotes the derivative of h, namely, the slope of the tangent line, at q'. Therefore convexity means that $h(q) \geq h(q_o) + h'(q_o)(q - q_o)$ for all $q_o > 0$. Now let $q_o = E(q)$ and establish (3.10), which can be restated more compactly as $E(h(q)) \geq h(E(q))$. Hint: recall the properties of a mean value of a nonnegative random variable.

3.6.7. Optimization of the citywide deployment of fire companies using the formula (3.17) guarantees a level of efficiency that slights other considerations of equity, as discussed in Section 3.5. A modification of (3.17) offers a partial way out of the dilemma. Replace each of the expressions $(N_i - \lambda_i/\mu_i)^{-1/2}$, which we rewrite as $g_i(N_i)$ for short, by $g_i(N_i)^b$, where b is a nonnegative parameter. When $b = 1$ one recovers (3.17), but if b is taken to be smaller than one, then each term in the sum becomes less dependent on response distance, especially as b gets closer to zero. On the other hand, choosing b greater than one the opposite occurs and, as b gets larger and larger, the terms with the larger travel distances begin to dominate the sum. Explain how the conflicting measures of performance, namely, efficiency of coverage, workload imbalance, and coverage imbalance, all discussed in Section 3.5, can be traded off against each other by letting b vary over nonnegative values. Having chosen a value of b the suitably modified form of (3.17) is then minimized as before.

3.6.8. The repositioning problem discussed in Section 3.5 can be written as an integer program, similar to those discussed in the previous chapter. Let $a_{i,j}$ equal one if the jth busy company belongs to the ith uncovered response neighborhood, and zero otherwise. Also, define x_j to be one if the empty firehouse of the jth busy company is to be temporarily filled by repositioning some other available unit. We wish to minimize the number of companies that must be relocated subject to the condition that no response neighborhoods remain uncovered. Express this mathematically.

3.6.9. A contagious disease (such as smallpox) has a lengthy incubation period so that a newly infected person manifests no symptoms for a period of time that we take to be exponentially distributed, with an average asymptomatic latency period of $1/\mu$. Each infected person who shows no symptoms can infect other susceptible individuals. The number of individuals infected this way is a counting process that is assumed to be Poisson at a rate λ that depends on the frequency of contact between susceptibles and infectives as well as the intimacy of contact. Even social customs may affect this rate as, for example, depending on whether people who meet embrace or simply nod to each other politely. There is a superficial resemblance here to the fire department problem in which the number of new infectives are arriving alarms and the incubation period is the service time of a busy fire company. Suppose that there are N sick people. Following the discussion in Section 3.2 show that the probability of exactly k new susceptibles becoming infected during the period in which no sick person shows any symptoms is $(\lambda/(\lambda+\mu))^k(\mu/(\lambda+\mu)$ with an average number of new infectives being λ/μ. Hint: recall the properties of the geometric distribution (Exercise 1.5.1).

3.6.10. Continuing with the previous exercise, let $\theta = \lambda/\mu$ and assume that $\theta < 1$. Each generation of infectives behaves in the same way as the previous one and if W_m denotes the number of infectives in the mth generation, then, because there are W_{m-1} infectives in the previous generation,

$$W_m = \sum_{i=1}^{W_{m-1}} U_i$$

where U_i is the number of cases transmitted by the ith infective. The expected value of U_i, regardless of m, is θ, as we saw in the previous exercise. That is, $E(U_i \mid W_{m-1}) = \theta$. Using the same result concerning conditional means that was employed in deriving (3.10), namely, that $E(W_m) = E(E(W_m \mid W_{m-1}))$, as discussed in the appendix, it follows that

$$E(W_m) = E\left(\sum_{i=1}^{W_{m-1}} \theta\right) = \theta E(W_{m-1})$$

Now $W_0 = 1$, because all infectives come from a single sick person in the zeroth generation. Therefore we can solve for W_m recursively. Using this, compute the total average number of infectives produced by the first sick person. Throughout these computations it has been assumed that once a sick person manifests symptoms, he or she is removed from society by quarantine or hospitalization and so no longer transmits the disease to anyone else.

Finally, establish that the epidemic eventually dies out by verifying that prob $(W_m = 0) \to 1$ as $m \to \infty$.

3.6.11. In this chapter we worked with response distance D rather than response time T. Both are random variables, and so is the speed of travel S of a fire engine. Because it is true that $T = D/S$, it was assumed that $E(T) = E(D)/E(S)$.

But in fact, $E(T) = E(D)E(1/S)$ and $1/S$ is a convex function of S. Using arguments similar to those in Exercise 3.6.6 show that actually $E(T) \geq E(D)/E(S)$.

3.6.12. We are given a region of area A as shown in the Figure 3.1, in which an incident is located at the center where the x and y coordinates are zero. A firehouse is distributed at random *uniformly* (not Poissonly!) within the region and a vehicle is dispatched from the firehouse along a right angle distance metric. A typical path is indicated by the arrows in the figure. Compute the mean travel distance $E(D)$ to the incident. Hint: the locus of points a distance r from the origin has area $2r^2$ and the probability that a firehouse is within a distance r of the incident is therefore $2r^2/A$.

3.6.13. Show that the variance of a Poisson random variable is equal to its expected value (see Exercise 3.5.1). This fact is used in the next exercise.

3.6.14. In experiments carried out by Luria and Delbruck in 1943, a large culture of bacteria, grown from a single cell over several generations (doubling in population with each generation), is exposed to a bacteriophage (virus). The virus attacks the bacteria and most are lysed. However, a few survive and give rise to colonies that are resistant to the virus and as each of these multiply they spawn bacteria that are also resistant. Evidently mutations occur in the bacteria that allow them to resist lysis. Examples of this phenomenon abound and include the well-known resistance of certain microorganisms to antibiotics. What is not immediately obvious is whether the mutations occur as a result of being exposed to the phage (virus) or if mutations occur spontaneously throughout the growth of the culture so that resistant colonies already exist when they are eventually exposed to the phage. The alternative is between a mutation that is an adaptation to an induced challenge to survival or if it occurs sporadically over time as a natural event prior to being treated with the phage. In other words, "is nature an editor or a composer?"

What Luria and Delbruck proposed is to generate a large number of independent cultures from a single cell and then count the number of resistant bacteria just after the growth medium has been impregnated with the harmful phage. If mutations occur only on contact with the phage, there is a small probability of survival due to a favorable mutation and the distribution of favorable mutations after many such experiments should be roughly Poisson. In this case, the mean and variance of the size of the surviving mutants would be nearly equal (Exercise 3.6.13).

On the other hand, if mutants occur haphazardly in each generation as the culture is growing, then why would you expect the variance to be much larger than the mean? In fact, this is what Luria and Delbruck actually found in their investigation, leading them to suggest that mutations are spontaneous events.

3.6.15. Consider a single server whose service time distribution is exponential and mean service rate μ, and suppose that the arrivals are Poisson at rate λ. Show that the departures are also Poisson at the same rate λ. This is a special case of what is known as Burke's theorem (see reference [16]). Hint: Let T be the time between departures and show that T has an exponential distribution at mean rate λ. This implies that the sequence of successive departures constitutes a Poisson process.

One consequence of this result is that if two servers operate independently but in tandem, then the output of the first server, namely, the input to the second server, is Poisson distributed at the same rate as the input to the first server.

3.7. Further Readings

There are a number of books on stochastic processes that discuss the Poisson distribution in detail, at a level commensurate with that adopted in this chapter. We can recommend the most recent edition of a book by Ross [46].

The inverse square root law for fire department operations was derived and tested by Kolesar and Blum [34]. A more comprehensive discussion of the models employed in this chapter, together with a thorough treatment of the Rand Fire Study, with further references, is in the book [17]. This includes the question of which units to deploy to a given alarm and the issue of how many to dispatch. The deployment of fire companies in multiple stages, the topic of Section 3.4, is due to Jan Chaiken and can be found in the report [18] and in Chapter 7 of the book [17].

As noted previously, the inverse square root law was a key tool in deciding how many fire companies to disband and how many to relocate during the fiscal crisis of New York City in 1971. This controversial move engendered protests at city hall and a suit in federal court by the firemen's union and by irate citizens. However, after the analysis was explained in court and at briefings to local groups, the opposition died down and the changes took place at a considerable saving to the city (see "Union Fights Fire Department Cuts," *New York Times*, December 22, 1972).

The Poisson distribution has wide applicability in the modeling of stochastic processes. This is illustrated in exercises 3.6.9 and 3.6.10, which discuss a model for the spread of an infectious disease. Further details on this topic are available in the little booklet [27] and in the book by Bailey [4]. A mathematical treatment of the Luria and Delbruck experiment of exercise 3.6.14 is contained in [36].

Clean Streets

4.1. Background

A truck with a large rotating mechanical broom attached to its front rumbles slowly down a residential street cautiously avoiding the few illegally parked cars that hinder its progress. After sweeping several miles of neighborhood streets, the truck returns, some hours later, to the depot it started from. Along the way it continues to sweep, on one side of a street or the other; if it has already passed in this direction it lifts the broom and moves on to the next street along the route that has yet to be swept. When this happens some valuable time is wasted; the unproductive travel is called deadheading and the goal is to find a route that requires the least amount of deadheading. A version of this problem was solved by the mathematician Leonhard Euler more than two centuries ago, as we discuss in Section 4.2.

Closely related to street sweeping and of greater economic significance is household refuse collection in which a truck and its crew pick up residential garbage from curby cans that are placed by the residents along the edge of the street curb. Since the truck must traverse the entire length of a street in order not to miss any residents or businesses along the way, its path from depot to dumpsite is not dissimilar to that of a sweeper in the need to keep deadheading to a minimum. These types of routing problems are discussed in Section 4.3.

In contrast to finding good routes, there is a question of how many vehicles to assign to carry out a stipulated pickup and/or delivery service, as in schoolbus routing or, again, refuse collection. The problem now is to determine the least number of buses, or trucks, as the case may be, that must be employed to complete the assigned routes within a specific time period and subject to capacity constraints. We take a brief look at this in Section 4.4 for the specific situation in which garbage trucks are scheduled for refuse collection while satisfying the daily and weekly demand for garbage pickup. This is based on a study done for New York City several decades ago.

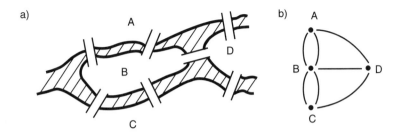

Figure 4.1. Konigsberg bridge problem. The shaded area is the river Pregel.

4.2. Euler Tour

When Leonhard Euler solved the puzzle of the Konigsberg bridges in 1743 he unwittingly uncovered a class of problems that were to have important applications several centuries later. The town of Konigsberg, as it was then known, was located at the confluence of two branches of the river Pregel in eastern Europe, as shown in Figure 4.1a. Note that several bridges link the mainland to the opposite banks and to the island, and the puzzle was to determine if a round-trip walking tour of the town could be carried out that passes over each of the seven bridges exactly once. Euler showed this to be impossible. To place the puzzle into a mathematical context, the map in Figure 4.1a is replaced by a schematic in which each of the four connecting land masses are represented by a node, labeled as A, B, C, and D, and the bridges by edges joining the nodes, as in Figure 4.1b. Any collection of nodes joined by edges is called a graph and the *Konigsberg puzzle* can be generalized, as Euler recognized, to ask whether a round-trip tour through any connected graph (that is, without disjoint subsets of nodes) can be found that covers each edge exactly once. Euler showed that such a tour is possible if and only if the number of edges incident to any node is even. His reasoning is discussed in a moment after a few preliminaries. Observe that the graph in Figure 4.1b violates this node condition and so the Konigsberg puzzle is not solvable.

Define the *degree* of a node in a graph to be the number of *edges* incident to that node and denote it by $\delta(v_i)$, where v_i indicates the ith node. The graph is said to be *directed* if each edge is assigned a specific direction. In this case, we also define the *inner or outer degree* of a node to be the number of edges directed into it or out of it, respectively, and indicate them as $\delta_+(v_i)$ and $\delta_-(v_i)$. Note that $\delta(v_i) = \delta_+(v_i) + \delta_-(v_i)$. Several easy lemmas are needed:

Lemma 4.1 $\sum \delta(v_i) = 2e$, *where e is the number of edges in the graph and the sum is taken over all nodes.*

Proof: This follows from the observation that each edge is incident to exactly two nodes. □

Lemma 4.2 *The sum of the degrees over all odd degree nodes is even.*

Proof: Let N_1 and N_2 denote the subsets of even and odd degree nodes. Then

$$\sum \delta(v_i) - \sum_{N_1} \delta(v_i) = \sum_{N_2} \delta(v_i)$$

The first term on the left, a sum over all nodes, is even because of Lemma 4.1. The second term is also even since it is a sum of even numbers. Therefore, the right side is also even. □

An easy consequence of Lemma 4.2 is the following, the proof of which is left as exercise 4.5.5.

Lemma 4.3 *There are an even number of odd degree nodes.*

A closed path in a finite graph (namely, a graph with a finite number of nodes and edges) that covers each edge exactly once is called a *Euler tour*, regardless of whether the graph is directed or not.

Theorem 4.1 *A closed path in a connected directed graph is a Euler tour if and only if the difference between the inner and outer degrees is zero at each node.*

Proof: If a Euler tour exists it uses one edge to arrive at a node and a different one to leave. It follows that $\delta_+ = \delta_-$ at each node.

Conversely, if the degree condition is satisfied, start a tour at some arbitrary node v_1 and move along a previously unused edge to an adjacent node. It is evident that each time we enter a given node that is different from v_1, it is also possible to leave along a different edge. Continue in this manner forming a path along adjacent nodes until a closed path M_1 is formed that terminates at some node v_2. If M_1 covers all the nodes we are done. Otherwise there must be an even number of unused edges incident to v_2. Remove all the edges in M_1 and begin anew starting now at v_2 using the same procedure to obtain a new closed path M_2. If M_1 and M_2 together exhaust all the nodes in the graph we are done. Otherwise repeat these steps until no further edges exist. Sooner or later this will happen since the graph contains a finite number of edges. □

A fairly immediate corollary to this theorem is Euler's original result, whose proof is left as exercise 4.5.1.

Theorem 4.2 *In a connected graph that is not directed a Euler tour exists if and only if the degree of each node is even.*

Another Lemma is needed later whose proof is quite easy and is obtained as exercise 4.5.3.

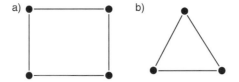

Figure 4.2. A bipartite graph (two-colorable) in panel a, but not in panel b.

Lemma 4.4   *If the degree condition of Theorem 4.2.1 holds, then*

$$\sum \delta_+(v_i) = \sum \delta_-(v_i)$$

where the sums are over all nodes in the graph. Moreover, $\sum \delta_+(v_i) = e$.

Before moving on to applications one more theorem is required. A *bipartite graph* is one in which all the nodes are divided into two classes such that no edge exists between adjacent nodes in the same subclass. Put another way, adjacent nodes have an edge between them only if they belong to different subclasses. In Figure 4.2a the graph is bipartite, but it is not so in Figure 4.2b.

An alternative way to express bipartiteness is to say that the graph is two-colorable in the sense that if each node is painted with either red or green then adjacent nodes have different colors. So, again the graph of Figure 4.2b is not two-colorable. The requirement of colorability is used in Section 4.4.

Theorem 4.3   *A connected undirected graph is two-colorable if and only if it contains no cycles (namely, a closed loop) consisting of an odd number of edges.*

Proof: If there are no odd-order cycles, pick a node and paint it with one of the two colors and then paint all nodes adjacent to it with the other color. Repeat this procedure until all nodes are painted. In this manner no node can be painted with both colors because this would imply that we can reach this node from the starting node along two different paths, one of which must have an even number of edges and the other an odd number. But this means that a cycle has been formed with an odd number of edges, which is a contradiction.

Conversely, if the graph is two-colorable the pigments must alternate when a cycle is traversed and this implies that the cycle must contain only an even number of edges.

Note that what is wrong with Figure 4.2b is that it is an odd-order cycle.   □

4.3. Street Sweeping

In New York City, as in other places, street sweeping with mechanical brooms is an elusive and hard way to quantify measure of cleanliness. Therefore, a surrogate is adopted of finding minimum time tours through the street network and, if more than one vehicle is required, of finding routes that utilize the least number of vehicles.

a) b)

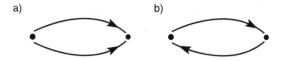

Figure 4.3. Directed edges to represent the two sides of a street for one-way (a) and two-way (b) streets.

Presumably these goals allow for the greatest number of streets to be swept in a given time period with a given number of vehicles.

The task of sweeping and, for that matter, of flushing the streets with water sprinkling trucks, requires that the vehicles move along the entire street network in a continuous roundtrip. If each street is covered exactly once without overlap this must be the minimum time path. Unfortunately, in practice it is generally true that some streets need to be traversed more than once if the network is to be fully covered. Streets that are traversed but not swept are said to be deadheaded, as was discussed in Section 4.1. The problem, then, is to find a roundtrip tour that minimizes the necessary amount of deadheading. This means finding a set of deadheading links of shortest possible length that, when added as edges to the original graph, corresponds to an augmented street network in which Euler's condition of Theorem 4.2.1 is satisfied. When a truck completes a cycle through the enlarged network the round-trip is optimal in the sense that even though some edges are duplicated the total added travel time or distance, as the case may be, is as small as possible. We will explore this question in the context of directed street networks, namely those that are represented by directed graphs.

The reason for directed networks is that some, if not most, streets are one-way and because in the process of sweeping or flushing each side of the street needs to be covered separately in the direction of traffic. For one-way streets this means that each side is to be swept once but in the same direction. Figure 4.3a shows the graphical version of this fact, using two edges to represent the sides of a one-way street that terminate at intersections denoted by the two nodes. Otherwise, for a two-way street, each side is swept in opposite directions (Figure 4.3b).

Before embarking on details let us note that essentially the same approach will be valid in a number of other municipal contexts. For instance, the spreading of sand and salt on icy roads entails a routing along each edge exactly once. Another case is that of household refuse collection in which there are sufficiently numerous pickup points on any given street that the garbage truck must necessarily traverse the entire street as if the aggregate refuse were distributed continuously along the edge. Deadheading is not a problem here since the trucks can move quickly when they aren't collecting trash. On the other hand, unlike the sweeper, the collection truck must go to a dumpsite to unload when its capacity is met and then return back to the streets. Fortunately these variants are readily accommodated within the procedure to be described presently.

Consider, then, a sample fragment from an urban street network, as shown in Figure 4.4 in which the arrows indicate whether or not the street is one-way.

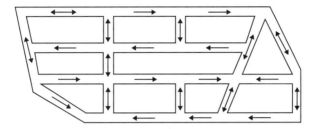

Figure 4.4. A portion of a street network. Arrows indicate one- and two-way streets.

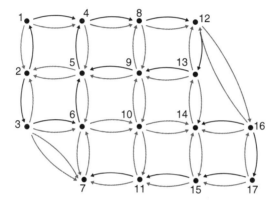

Figure 4.5. Graphical representation of the preceding street network. Dark lines indicate no parking is in effect between 8 and 9 A.M., while the lighter lines mean that there are no parking restrictions during that time period.

Assume that no parking regulations are in effect from 8 A.M. to 9 A.M. on certain sides of several streets. This is indicated in Figure 4.5, which is a directed graph representation of the network, by heavy dark lines to signal those curbsides that can be swept during this 1-hour window. The lighter lines denote street segments that are not available for sweeping but that can be used for deadheading.

If we extract from this graph the subset of edges that are to be covered by a Euler tour the directed graph of Figure 4.6 results. This entails a routing problem that consists of 17 nodes and 16 edges in which we note that some nodes have more edges entering than leaving and vice versa. It follows from Theorem 4.2.1 that some deadheading will be required and the optimization program to be formulated next is designed to find a set of duplicate edges of shortest total length.

To do this we define the *polarity* of a node as the difference between the number of edges that enter and those that leave. For a Euler tour, all nodes must have polarity zero, as we know, but after inserting an appropriate number of deadheading links into the graph in Figure 4.6, each node of the augmented graph will have zero

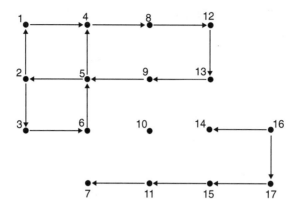

Figure 4.6. Subgraph of the previous Figure 4.5 indicating those streets that can be swept from 8 to 9 A.M.

polarity, as required. The nodes having nonzero polarity are:

Node number:	4	7	14	2	16
Polarity	1	1	1	−1	−2

Let us agree to denote all nodes with positive polarity as *supply* nodes because they have an excess of incoming edges and are therefore in a position to donate some outgoing ones. Similarly we designate as *demand* nodes those with negative polarity because they are in a position to be edge recipients. It should be evident that to say that a node has positive polarity is tantamount to asserting its inner minus outer degree is positive and that negative polarity is equivalent to inner minus outer degree is negative. That is, polarity equals $\delta_+ - \delta_-$ at each node. It follows directly from Lemma 4.4 that the sum, over all nodes, of the supplies (positive polarities) equals the sum of demands (negative polarities) and this is the key to the procedure to be described next.

Each supply node i can be linked to a demand node j by a set of edges and this is to be done in a least cost manner, where cost is taken to be the travel time c_{ij} between them. The total demands will exactly match the total supplies, as we know. In Figure 4.6 total demand and total supply are each 3.

List the supply nodes 4, 7, and 11 as s_1, s_2, and s_3 and the demand nodes 2 and 16 as d_1 and d_2. Next, we need to determine the travel times c_{ij} along shortest paths joining i to j. It is important to recognize that this path can be taken along any street in the original network of Figure 4.4. There are formal algorithms for computing shortest paths that one can use but in our simple example this can be done by visual inspection. Assume, for simplicity, that crosstown travel times are 8 min per block and are 5 min for each up- or downtown block. Two exceptions are the links $12 \rightarrow 16$ (10 min) and $3 \rightarrow 7$ (8 min). For these values, we see that

Table 4.1. Travel times
between supply and demand
nodes in Figure 4.5.

	d_1	d_2
s_1	13	26
s_2	18	41
s_3	29	20

the shortest path from s_2 to d_1 is given by 7→6→5→2 for a total of 18 min. It is left to you to complete the travel times for the remaining pairs of nodes (Exercise 4.5.4). This leads to Table 4.1.

It is now possible to pose an integer programming problem to minimize the total time required to join the supply and demand nodes, subject to satisfying the supply and demand at each node. Let $x_{ij} \geq 0$ denote the number of times that supply node i is to be linked to demand node j along the shortest path. Then the problem is to minimize

$$\{13x_{11} + 26x_{12} + 18x_{21} + 41x_{22} + 29x_{31} + 20x_{32}\}$$

subject to

$$
\begin{aligned}
x_{11} + x_{12} &&&&&&= 1 \\
&& x_{21} + x_{22} &&&&= 1 \\
&&&& x_{31} + x_{32} &&= 1 \\
x_{11} + && x_{21} + && x_{31} &&= 1 \\
x_{12} + && x_{22} + && x_{32} &&= 2
\end{aligned}
$$

As is readily verified the optimum values of the assignments are $x_{11} = x_{22} = x_{31} = 0$, $x_{12} = x_{21} = x_{32} = 1$. This means that s_1 is to be joined to d_2, s_2 to d_1, and s_3 to d_2, each once. By adding these additional paths to the network of Figure 4.6 we obtain the augmented graph of Figure 4.7 in which the supplementary links are indicated by dotted lines. The new edges represent streets that must be deadheaded and the optimization assures us that this is the least amount of extra travel that we can get away with. Note that the polarity of each node in Figure 4.7 is now zero and therefore a Euler tour can be constructed. Simply begin anywhere and traverse a single closed loop. For example, the following cycle is a Euler tour:

$$1 \rightarrow 4 \rightarrow 8 \rightarrow 12 \rightarrow 13 \rightarrow 9 \rightarrow 5 \rightarrow 4 \rightarrow 8 \rightarrow 12 \rightarrow 16 \rightarrow 14 \rightarrow 13 \rightarrow 12$$
$$\rightarrow 16 \rightarrow 17 \rightarrow 15 \rightarrow 11 \rightarrow 7 \rightarrow 6 \rightarrow 5 \rightarrow 2 \rightarrow 3 \rightarrow 6 \rightarrow 5 \rightarrow 2 \rightarrow 1$$

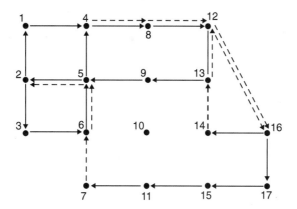

Figure 4.7. Euler tour for street sweeping from 8 to 9 A.M. Dotted lines indicate dead-heading.

We are not yet through. Because the parking regulations last 1 hr it is not possible for a single truck to cover the entire network and so the path must be partitioned into segments, each of which is less than an hour's duration. But this gives us an opportunity to reduce deadheading even further, as we will see. It is typical of problems of this type that the time required for sweeping or collecting refuse lies well within the time available in a workshift and therefore time to travel to and from a depot to the starting point of a tour can be treated as incidental and something that can be ignored. One may assume that each truck is positioned to begin its route at any point along the network at the beginning of the 8 A.M. starting period and this permits certain deadheading segments to be eliminated at either the front or tail end of a tour. This can be illustrated by breaking the Euler tour already constructed into three truck routes with the elimination of 44 min of unnecessary travel:

Route 1: $1 \rightarrow 4 \rightarrow 8 \rightarrow 12 \rightarrow 13 \rightarrow 9 \rightarrow 5 \rightarrow 4$ (50 min)

Route 2: $16 \rightarrow 14 \rightarrow 13 \rightarrow 12 \rightarrow 16 \rightarrow 17 \rightarrow 15 \rightarrow 11 \rightarrow 7$ (57 min)

Route 3: $2 \rightarrow 3 \rightarrow 6 \rightarrow 5 \rightarrow 2 \rightarrow 1$ (31 min)

The final set of tours covers the entire network using only 20 min of deadheading, whereas the total sweeptime is 118 min. Indeed, only route 2 requires any overlap of its own path.

This example serves as a demonstration of how edge-routing problems can be handled in general. We summarize its salient features by listing the main steps of the analysis. First, construct a directed graph G to represent the entire street network and extract a subgraph G_1 to denote those streets that, because of parking regulations, are available to be swept (or have refuse collected, as the case may be) during a given time period. Then isolate all nodes in G_1 that have nonzero polarity and label them either as supplies s_i or demands d_j. Compute the shortest path in G

(that is, using any edge in the original graph) between the supplies and demands and solve the problem of finding the smallest total travel time between supply and demand. That is,

$$\text{minimize} \sum \sum c_{ij}\, x_{ij}$$

subject to

$$\sum x_{ij} = s_i, \quad \text{with the sum over } j$$
$$\sum x_{ij} = d_j, \quad \text{with the sum over } i$$

and $x_{ij} \geq 0$. The values of x_{ij} indicate the number of deadheading links to insert between s_i and d_j. Augment G_1 to a graph G_2 in which all the nodes now have zero polarity by adding in these extra links and then form a Euler tour. Finally, break the tour up into individual truck routes, each satisfying a constraint on the time available for sweeping or collecting.

In closing, it is important to note that the routes formed by the mathematical scheme outlined here may not be implementable in practice. The reason is the trucks sometimes find it inconvenient to maneuver from one side of a street to another and that awkward turns, such as U-turns, or frequent changes from sweeping to deadheading and back, may foul up drivers who in general prefer long straight segments of the network whenever possible. This again illustrates the pitfall of not doing analysis in the context of an operating environment. Some changes may cost more to implement than they are designed to save. For example, better routes could conceivably be obtained by altering the direction of certain streets or by changing the parking regulations. But the expense of sign changes and the disruption of traffic patterns that ensue when regulations are tampered with may well preclude such "solutions."

4.4. Vehicle Scheduling

A city has a fleet of large trucks that pick up refuse at selected nonresidential sites such as schools, hospitals, and other public buildings where there are trash bins that are hoisted onto the truck by fork lift and then compacted. When the trucks fill up, they go to a disposal site to empty their load and then they return to pick up more refuse at other locations. In this way, they make a number of back-and-forth trips that can be linked together to form a daily schedule for a single truck, provided that the total time does not exceed the time allotted for a single work shift (a workday of 6 to 8 hr). The concatenation of several back-and-forth tours to a dump to form a daily route that satisfies the time constraint is called a *feasible truck schedule*. Implicit in this that each tour requires less time to complete than the time available in a shift, in conformity with the usual situation in which time is the binding constraint in forming daily routes out of tours, whereas vehicle capacity is the limiting factor in the formation of the tours themselves. Generally more than one vehicle is required to satisfy all the clients in 1 day and so one wants to form schedules that require the least number of trucks.

Suppose that a large number M of such feasible truck schedules have been put together from the separate round trip tours that cover all the clients, at a total of N locations. This is usually done by "eyeballing" a street map of the region being serviced, although more sophisticated techniques are available to do this such as the one discussed in the preceding section. There may be considerable overlap among the M schedules since there are several ways of hooking together round-trips to the dump without exceeding the time constraint.

Since we want a selection of schedules that uses the least number of trucks, this can be posed as an optimization problem. In the spirit of the districting problem of Chapter 2, let $a_{i,j} = 1$ if the ith client belongs to the jth schedule, with $a_{i,j} = 0$ otherwise. Also, let $x_j = 1$ if schedule j is selected, with $x_j = 0$ otherwise. Then one wishes to minimize the sum

$$\sum_{j=1}^{M} x_j$$

subject to the condition that each pickup point lies on some route or other:

$$\sum_{j=1}^{M} a_{i,j} x_j \geq 1$$

for $i = 1, 2, \ldots, N$. The last inequality ensures that all N pickup points are serviced.

A somewhat different set of scheduling problems arise from the fact that in practice not all sites are serviced the same number of times each week. Up to now we assumed that the daily and weekly schedules for a truck are the same, but in some areas the clients have their refuse removed on Monday, Wednesday, and Friday (MWF), others on Tuesday, Thursday, and Saturday (TTS), while the remainder are serviced every day (except Sunday). In this context, the formation of a weekly schedule is decoupled from that of finding daily schedules. If one assumes that three-times-a-week customers are indifferent to whether they get a MWF or TTS schedule, the problem is to assign tours to days of the week so as to minimize the number of required vehicles. Rather than treat this problem directly we focus instead on the subtle question of whether weekly schedules can be formed at all.

We begin by collecting a bunch of roundtrip tours to the dump, much as before, except that we label them as either "red" or "green" tours, corresponding to MWF or TTS. The only proviso is that six-times-a-week pickup points must appear on both a green and red tour if they are to be serviced once every day but Sunday, and not twice on the same three days of the week. However there may be a difficulty here. To see how this can arise let us form a *tour graph* in which the nodes denote tours, and an edge is inserted between two nodes whenever the corresponding tours have a six-day-a-week point in common. Recall that a graph G is a set of points called nodes that are joined by line segments called edges.

The six-times-a-week sites are to appear on exactly two tours having distinctly different colors, and so adjacent nodes in the tour graph must be painted differently. That this cannot always be done is shown by the graph in Figure 4.2b, where we see that the service frequency requirement is violated. In this case the set of tours is infeasible.

As discussed in Section 4.2, the least number of colors necessary to paint the nodes of any graph so that no adjacent node has the same color is called the chromatic number of the graph and we saw there that a tour graph is two-colorable if and only if it has no cycles of odd order.

The trouble with the graph depicted in Figure 4.2b is that it consists of a single cycle of odd length. In general the same difficulty can arise with other service frequency requirements. Consider, for example, sites that are serviced either Monday–Thursday, or Tuesday–Friday, or Wednesday–Saturday, or every day of the workweek. In this case, the round-trip tours are painted with three colors for the three different service frequencies, with six-times-a-week points to appear exactly once on a tour of each type to guarantee that they be serviced once every day of the week and not several times on some subset of the days.

Form a tour graph as before, with an edge between nodes if the tours have a six-day point in common, and label all nodes as either red, green, or blue. It is apparent that the service frequency requirement can be met in a feasible manner only if the tour graph can be painted with the three colors in such a way that adjacent nodes have different pigments. That is, we want the graph to have a chromatic number of three. Figure 4.8 shows that in general this cannot be done.

A question of interest is to determine the chromatic number of a graph. To do this one defines an *independent set* in a graph to be a subset of nodes such that no two of them are adjacent. Index all maximally independent sets by $j = 1, 2, \ldots, M$, and label them by I_j. Let $a_{i,j} = 1$ if node i belongs to I_j, with zero otherwise, and let $x_j = 1$ if subset I_j is chosen, with zero otherwise. Then one wishes to minimize the number of independent sets that are required to cover the graph completely. Some of the I_j may overlap, of course. The mathematical formulation of this

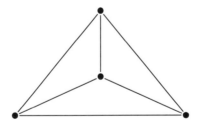

Figure 4.8. A tour graph that is not three-colorable.

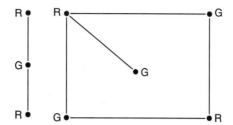

Figure 4.9. A tour graph with different tour assignments indicated as R (red) and G (green) tours.

optimization problem is left to Exercise 4.5.7, and its solution tells us whether the collection of given tours is feasible in terms of satisfying the service frequency.

With multiple service frequencies the problem of finding the least number of garbage trucks is now more complicated. Assuming that a feasible coloring exists one needs to form daily routes that pick up all points on MWF or TTS, say, while satisfying the time constraint. Consider, for example, the tour graph shown in Figure 4.9. The graph is not connected since there is no a priori reason why round-trip tours must share the same six-times-a-week point in common. Because there are no circuits of odd length, Theorem 4.3 tells us that it can be two-colored. There are evidently four ways of coloring the graph, two of which are indicated in the figure. Assuming that each truck can cover only two tours a day, the assignment of red or green can make the difference between needing two trucks every day or three on alternate days, and so the task of finding the least number of vehicles must now take the choice of coloring into consideration. However, this will not be pursued further.

4.5. Exercises

4.5.1. Establish Theorem 4.2 as a consequence of Theorem 4.1.

4.5.2. In Figure 4.1b suppose that the travel times (in minutes) between nodes are, in either direction, 5 from A to B, 7 from B to C, 10 from D to A or D to B, and 5 from D to C. We know from Theorem 4.2 that a Euler tour does not exist. Add deadheading links to this graph so that a roundtrip path starting from D requires the least amount of time and indicate the ordering of the edges in this round-trip.

4.5.3. Prove Lemma 4.4. Hint: use Lemma 4.1.

4.5.4. Find the shortest paths between supply and demand nodes to fill in Table 4.1.

4.5.5. Prove Lemma 4.3.

4.5.6. A procedure for finding deadheading links for directed graphs, as discussed in Section 4.3, utilized the fact (Lemma 4.4) that the sum of degrees of outgoing and incoming edges is the same. For undirected graphs, as in Euler's Konigsberg problem, a similar procedure is possible because of Lemma 4.3. Discuss what such a procedure would be.

4.5.7. Complete the formulation of the integer programming problem of finding the chromatic number of a graph, as discussed in Section 4.4.

4.6. Further Readings

Vehicle routing and scheduling are treated in the papers by Beltrami and Bodin [11] and Tucker [49] and these are based, in part, on experiences in advising the New York City Sanitation Department.

The Coil of Life

5.1. Background

Although the human genome has now been decoded, the focus of investigations on the molecular form and biochemical functions of the gene continues unabated. One question of interest has been the manner in which the long strands of DNA are packed into the tiny confines of a cell. This is a complicated topic, but we can obtain a brief glimpse into some of the mathematical issues it raises.

In the next section a formula is derived, due to Carl Friedrich Gauss, that computes the number of times two space curves link each other. This result is not only of interest in itself, and a good exercise in the multivariate calculus, but it also provides some of the background material needed for the introductory and somewhat intuitive discussion of DNA in Section 5.3. The idea there is to study the possible deformations of this molecule within a cell.

5.2. The Gauss Linking Number

Lines emanating from a point p that intersect a segment of a smooth curve in a plane subtend an angle $\theta = s/R$ on a circle of radius R having p as its center, where s is the arc length of the segment of the circle that is intersected by the lines (Figure 5.1). This can be extended to three dimensions to form a solid angle in the following manner. All the lines that emanate from a point p in space and intersect a segment S of smooth surface form a cone that intersects a sphere of radius L in a patch P of area $A(P)$ (Figure 5.1). The *solid angle* with vertex p subtended by S is defined to be $\Omega(p) = A(P)/L^2$.

Let $\mathbf{r}$ denote the position vector from p to any point of the surface S. In terms of coordinates, if $p = (a, b, c)$, then $\mathbf{r} = (x - a, y - b, z - c)$ and we define γ as the scalar product $r \cdot r = (x - a)^2 + (y - b)^2 + (z - c)^2$.

79

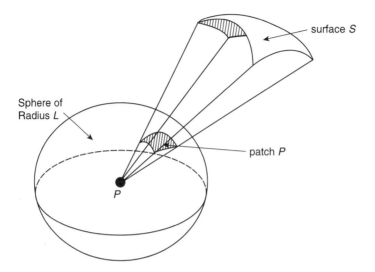

Figure 5.1. Solid angle $\Omega(p)$ with vertex p subtended by a surface S.

Consider the closed region R enclosed by S, P, and the surface of the cone between P and S (Figure 5.1) and let $\mathbf{n}$ denote the unit outward normal to R. Define a vector field $\mathbf{v}$ by $\mathbf{r}/\gamma^{3/2}$. Then the divergence theorem of the vector calculus tells us that

$$\iiint\limits_{R} \text{Div } \mathbf{v} \, dR = \iint\limits_{\partial R} \mathbf{v} \cdot \mathbf{n} \, dA$$

where ∂R indicates the boundary of R consisting of $S \cup P \cup$ boundary of the cone. It is a straightforward computation (Exercise 5.4.1) to show that Div $\mathbf{v} = 0$. Since $\mathbf{n}$ is orthogonal to $\mathbf{r}$, and therefore to $\mathbf{v}$, along the boundary of the cone, the double integral over this part of the boundary of R vanishes. Moreover, $\mathbf{n}$ is collinear to $\mathbf{v}$ but points in the opposite direction on P where $\gamma = L^2$. It follows that

$$0 = \frac{-1}{L^2} \iint\limits_{P} dA + \iint\limits_{S} \mathbf{v} \cdot \mathbf{n} \, dA \qquad (5.1)$$

The left side of this expression is simply the solid angle $\Omega(p)$, and so

$$\Omega(p) = \iint\limits_{S} \mathbf{v} \cdot \mathbf{n} \, dA \qquad (5.2)$$

An immediate consequence of (5.2) is that the solid angle does not depend on the radius L of the sphere and so we can conveniently choose the radius to be one from now on.

Let $\partial(S)$ denote the bounding curve of the surface segment S and suppose that S' is any other smooth surface segment having the same boundary $\partial(S)$. Call R' the region enclosed by $S \cup S'$. If p is outside R', then the same argument that led to (5.1), using the divergence theorem, now shows that

$$\iiint_{R'} \text{Div } \mathbf{v} \, dV = 0 = -\iint_S \mathbf{v} \cdot \mathbf{n} \, dA + \iint_{S'} \mathbf{v} \cdot \mathbf{n} \, dA \qquad (5.3)$$

and, therefore, in view of (5.2), the solid angle depends only on the boundary of a surface and not on the surface segment itself. Moreover, because $\Omega(p)$ is the double integral of $\mathbf{v} \cdot \mathbf{n}$ over R', and because S and S' are arbitrarily chosen surfaces, (5.3) shows that $\Omega(p) = 0$ for p outside any closed surface which, in this case, is $\partial R'$.

On the other hand, if p is inside R' choose a small sphere S_p about p entirely enclosed within R' and consider the annular region R'' lying between the outside of S_p and the boundary of R'. The previous argument now shows that

$$\iiint_{R''} \text{Div } \mathbf{v} \, dV = 0 = \iint_{\partial R'} \mathbf{v} \cdot \mathbf{n} \, dA - \iint_{S_p} \mathbf{v} \cdot \mathbf{n} \, dA$$

The integral of $\mathbf{v} \cdot \mathbf{n}$ on the right reduces to $1/\gamma \iint_{S_p} dA$ because $\mathbf{v} = \mathbf{r}/\gamma^{3/2} = \mathbf{n}/\gamma$ (note that $\mathbf{v}$ and $\mathbf{n}$ are collinear on S_p). But the double integral equals the area of S_p, namely $4\pi\gamma$, from which it follows that the integral of $\mathbf{v} \cdot \mathbf{n}$ over the boundary of R' is 4π. Thus $\Omega(p) = 4\pi$ whenever p is inside a closed surface, which, in this case, is $\partial R'$. This fact will prove useful momentarily.

We leave it to the somewhat tedious Exercise 5.4.2 to establish that

$$\text{Grad } \Omega(p) = \int_{\partial(S)} \mathbf{v} \times d\mathbf{r} = \int \left(\mathbf{v} \times \frac{d\mathbf{r}}{ds} \right) ds \qquad (5.4)$$

where Grad means "gradient of," $\mathbf{v} \times d\mathbf{r}/ds$ denotes the cross product of the two vectors, and the integral on the right is a line integral with respect to arc length s over the bounding curve ∂S.

We now come to our main point.

Let C_1 and C_2 be two smooth closed curves that link each other. We want to count the number of times they are entwined.

Suppose that C_1 bounds a surface S and let n_1 be the number of times C_2 pierces S in the direction shown in Figure 5.2a (for simplicity, think of C_2 as lying in the x, y plane, bounding a portion of the plane, with C_2 crossing this plane from $z > 0$ to $z < 0$). In the same manner, let n_2 be the number of times C_2 crosses from $z < 0$ to $z > 0$. The *linking number* $\text{Lk}(C_1, C_2)$ is defined as $n_1 - n_2$.

In Figure 5.2a the linking number is $+1$ because $n_1 = 1, n_2 = 0$, whereas the linking number of Figure 5.2b is 0 because n_1 and n_2 are both 1. Finally, in

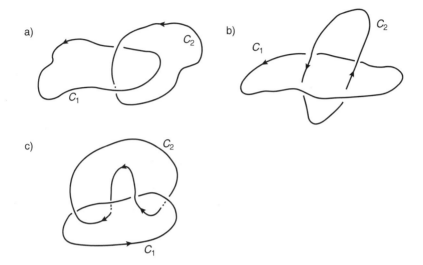

Figure 5.2. Three examples of the linking of two closed curves in space.

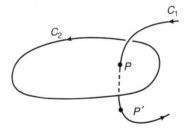

Figure 5.3. The linking of two curves in the vicinity of the surface S bound by one of the curves.

Figure 5.2c, the linking number is -2 because $n_1 = 0$ and $n_2 = 2$. We now show how $\text{Lk}(C_1, C_2)$ can be expressed as an integral.

It is fairly evident that $\Omega(p)$ changes in a continuous manner as p moves along C_1 and that the only source of discontinuity would occur when p penetrates S. To see what happens attach a second surface S' to the same boundary C_2 so that $S \cup S'$ enclose a region R. Let p' be some other point on C_1 that lies on a different side of S than does p and that is outside R, whereas p lies inside (Figure 5.3). As before, $\mathbf{n}$ denotes the outward unit normal to the closed surface $S \cup S'$.

Using the divergence theorem, together with relation (5.2), it was seen above that $\Omega(p) = 4\pi$, whereas $\Omega(p') = 0$ for the closed surface $\partial R = S \cup S'$. The outward normal $\mathbf{n}$ points in the opposite direction to the normal of the surface S

as defined by the orientation of its boundary C_2 and, therefore,

$$4\pi = \Omega_{\partial R}(p) = -\Omega_S(p) + \Omega_{S'}(p)$$
$$0 = \Omega_{\partial R}(p') = -\Omega_S(p') + \Omega_{S'}(p')$$

where the subscripts indicate the portion of solid angle generated by either ∂R, S, or S'. Subtracting these two relations gives

$$4\pi = \Omega_S(p') - \Omega_S(p) - \Omega_{S'}(p') + \Omega_{S'}(p) \tag{5.5}$$

Observe from (5.2) that the solid angle can be positive or negative depending on whether the angle from $\mathbf{v}$ to $\mathbf{n}$ is acute or obtuse. Because p and p' lie on the same side of S', by the way we positioned them (Figure 5.3) the last two terms in (5.5) have the same sign; therefore they cancel in the limit as $p - p'$ shrinks to zero. From (5.5) it now follows that as p and p' approach S from above and below, respectively, so that $p - p'$ goes to zero, $\Omega(p)$ jumps in value by 4π when it crosses the boundary.

Each time C_2 loops through C_1 by piercing S from above the solid angle increases by 4π and it decreases by 4π when it pierces S from below. It follows that the total change in Ω as p traverses C_1 is a multiple of 4π:

$$\int_{C_1} d\Omega = \int \frac{d\Omega}{d\tau} d\tau = 4\pi(n_1 - n_2) \tag{5.6}$$

where the integral on the right is a line integral with respect to arc length that we indicate by τ. The point p that traverses C_1 is parametrized by a position vector $\mathbf{r}_1(\tau)$, with parameter τ. Therefore, using the chain rule of differentiation, $d\Omega/d\tau = \mathrm{grad}\,\Omega(p) \cdot d\mathbf{r}_1/d\tau$ or $d\Omega = \mathrm{grad}\,\Omega \cdot d\mathbf{r}_1$. Employing (5.4) finally shows that

$$\int_{C_1} d\Omega = \int_{C_1} \mathrm{grad}\, d\Omega \cdot d\mathbf{r}_1 = \int_{C_1}\int_{C_2} (\mathbf{v} \times d\mathbf{r}_2) \cdot d\mathbf{r}_1 = 4\pi(n_1 - n_2)$$

The last expression can be written as $\int (\int (\mathbf{v} \times d\mathbf{r}_2/ds)\, ds) \cdot (\frac{d\mathbf{r}_1}{d\tau})\, d\tau$ for purposes of computation, where $\mathbf{r}_2(s)$ is a parametrization of C_2 with respect to arc length s (we use s here to distinguish the representation of the two curves).

Now $\mathbf{v} = \mathbf{r}/\gamma^{3/2}$ with $\mathbf{r} = \mathbf{r}_2 - \mathbf{r}_1$ being a vector directed from a generic position p on C_1 to some other point on C_2, and the length of $\mathbf{r}$ is $\gamma^{1/2}$. The vectors $d\mathbf{r}_1/d\tau$ and $d\mathbf{r}_2/ds$ are tangents to C_1 and C_2 and can be written as $\mathbf{T}_1$ and $\mathbf{T}_2$ respectively. Noting that $(\mathbf{v} \times \mathbf{T}_2) \cdot \mathbf{T}_1 = \mathbf{v} \cdot (\mathbf{T}_2 \times \mathbf{T}_1)$ gives us the expression we are looking for, namely,

$$\frac{1}{4\pi} \iint_{C_1 C_2} \mathbf{v} \cdot (d\mathbf{r}_2 \times d\mathbf{r}_1) = \frac{1}{4\pi} \iint \mathbf{v} \cdot (\mathbf{T}_2 \times \mathbf{T}_1)\, ds\, d\tau = n_1 - n_2 \qquad (5.7)$$

Expression (5.7) is known as *Gauss's linking formula* or, simply, as *Gauss's integral* and is often written as $\mathrm{Lk}(C_1, C_2)$. It is evidently always an integer, positive or negative. Although it is fairly clear on intuitive grounds that $\mathrm{Lk}(C_1, C_2) = \mathrm{Lk}(C_2, C_1)$, this also follows from (5.7) because a reversal of the roles of the two curves also reverses the sign of $\mathbf{v}$ and, moreover, $\mathbf{T}_2 \times \mathbf{T}_1 = -\mathbf{T}_1 \times \mathbf{T}_2$ so that integral is unchanged.

The derivation of (5.7) did not depend on how C_1 and C_2 are deformed in space as long as neither is allowed to intersect the other during a deformation. This means that the linking number is a topological invariant.

In Figure 5.2a the linking number is $+1$. It is useful to verify this directly by use of (5.7) (Exercise 5.4.3).

The Gauss linking formula will not be used explicitly in what follows because the linking number will be clear from the applications. However, its derivation is a prototype of the kind of mathematics that is needed for more profound investigations into the geometry and topology of DNA.

5.3. Twisting and Writhing of DNA

Without getting into unneeded details, DNA consists of four bases, labeled A, C, G, and T in which A pairs with T and C with G, each of these hooked together by a weak chemical bond. The pairs form nearly flat molecules that are stacked up like steps on a helical ladder. The edges of the bases are joined to two sugar-phosphate molecular strands that wind around the outside to form a double helix (Figure 5.4a). A simple model of the double helix represents the outer strands as two space curves C_1 and C_2 that form the bounding edges of a fictional ribbon that forms a continuous spiral "ladder" (Fig 5.4b).

The central axis of the ribbon can writhe in space, the two strands forming the boundary of the ribbon can coil about each other, and the whole ribbon can twist about its axis.

Some DNA occurs in nature as closed loops, and this is the configuration that will be considered here. The two strands defined by curves C_1 and C_2 link each other as they coil in and out giving rise to the linking number $\mathrm{Lk}(C_2, C_1)$ considered in the previous section. We now want to discuss twisting and writhing, beginning with the first.

Twisting is essentially a measure of the extent to which C_2 spins about C_1. It is probably more appropriate in the following to think of C_1 as the central backbone of bases in DNA rather than as the other sugar-phosphate strand, at least when talking about twist, but this distinction is sometimes blurred to simplify the discussion.

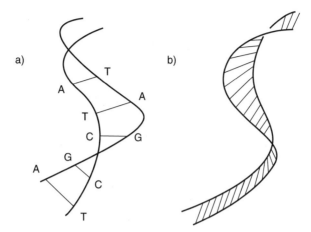

Figure 5.4. Schematic of DNA of two sugar-phosphate strands winding about a central backbone of bases (a) and a ribbon model of the DNA in which the edges can, depending on context, represent either two sugar-phosphate strands or one strand connected to the central backbone of bases (b).

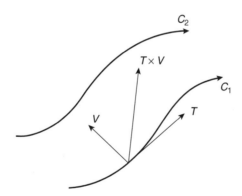

Figure 5.5. Local geometry of a strand twisting about a central backbone.

To be mathematically correct, assume that surface defined by the ribbon and its bounding curves are sufficiently smooth that all derivatives exist and are continuous. Let $\mathbf{T}$ denote the unit tangent vector to C_1 at any point s along this curve and let $\mathbf{V}$ be a unit vector orthogonal to $\mathbf{T}$ and directed toward C_2, tangent to the ribbon-like surface that stretches between the curves. Then $\mathbf{T}$ and $\mathbf{V}$ span the tangent plane to the ribbon at s and the cross product $\mathbf{T} \times \mathbf{V}$ is a unit vector orthogonal to the ribbon (Figure 5.5).

The vectors $\mathbf{T}$, $\mathbf{V}$, and $\mathbf{T} \times \mathbf{V}$ form an orthogonal frame that moves with s along C_1. The derivative $d\mathbf{V}/ds$ is the rate of change of $\mathbf{V}$ along C_1 and its component in the direction $\mathbf{T} \times \mathbf{V}$ is the scalar product $(\mathbf{T} \times \mathbf{V}) \cdot d\mathbf{V}/ds$.

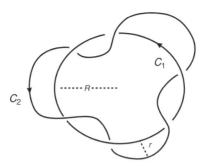

Figure 5.6. A helical curve winding about a circular backbone with three complete turns.

We now define the *twist* $Tw(C_2, C_1)$ of C_2 about C_1 as the total change in $\mathbf{V}$ in the direction $\mathbf{T} \times \mathbf{V}$ as the curve C_1 is traversed:

$$Tw(C_2, C_1) = \frac{1}{2\pi} \int_0^L \left[(\mathbf{T} \times \mathbf{V}) \cdot \frac{d\mathbf{V}}{ds} \right] ds = \frac{1}{2\pi} \int_{C_1} (\mathbf{T} \times \mathbf{V}) \cdot d\mathbf{V} \quad (5.8)$$

where L is the length of C_1. The factor 2π is included to express twist in number of turns rather than in angular change.

Note that twist is not an integer in general and its value depends on how the curves are deformed in space, as we show later. Also, the twist of C_2 about C_1 is not necessarily the same as the twist of C_1 about C_2 although we do not show this here. Unlike the linking number, therefore, twist is not a topological invariant.

Before moving on to the concept of writhing, it is useful to carry out the computation of twist in a specific case.

Choose C_1 to be a circle of radius R and length $L = 2\pi R$. Construct a torus (inner tube, if you wish) of radius r smaller than R having C_1 as its central axis. Let C_2 be a helical curve that lies on the surface of the tube so that it winds about C_1 n times. Figure 5.6 illustrates the configuration of the two curves in the case of $n = 3$.

Assume, more generally, that C_1 is a smooth closed curve lying in a plane that does not cross itself and defined, in terms of arclength s, by the vector function $\mathbf{x}(s)$. Moreover, let $\mathbf{y}(s)$ define a helical curve C_2 that wraps around C_1 on a tube of radius r centered at $\mathbf{x}(s)$. Then

$$\mathbf{y}(s) = \mathbf{x}(s) + r\mathbf{V}(s)$$

where $\mathbf{V}(s)$ is a unit vector, to be defined momentarily, that is directed from a point on C_1 to a corresponding point of C_2.

Take a cross-sectional slice of the torus by passing a plane through $\mathbf{x}(s)$, orthogonal to the unit tangent vector $\mathbf{T}$ to C_1 at $\mathbf{x}(s)$. Choose $\mathbf{N}$ to be a unit normal to C_1 at $x(s)$, lying in the plane of the curve, and let $\mathbf{B}$ denote $\mathbf{T} \times \mathbf{N}$. Then $\mathbf{B}$ is a

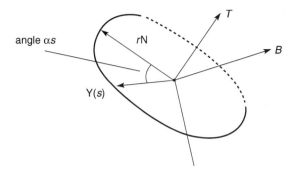

Figure 5.7. A plane slicing a torus whose central axis is a curve that is intersected orthogonally by the plane.

unit vector orthogonal to the plane containing C_1; $\mathbf{N}$ and $\mathbf{B}$ together determine the slicing plane (Exercise 5.4.4).

Define the unit vector $\mathbf{V}(s)$ by $\cos(\alpha s)\mathbf{N} + \sin(\alpha s)\mathbf{B}$, where $\alpha = 2n/L$. Then $\mathbf{y}(s) = \mathbf{x}(s) + r\mathbf{V}(s)$ is indeed a point on C_2; the quantity αs measures the angle from $\mathbf{N}$ to $\mathbf{y}(s)$ (Figure 5.7). There is, therefore, a one-to-one correspondence between C_1 and C_2 in which a point on C_1 defined by $\mathbf{x}(s)$ is uniquely identified with a point on C_2 defined by $\mathbf{y}(s)$ for each value of s from 0 to L. As s moves along C_1, the vector $\mathbf{y}(s)$ rotates about C_1 and generates the helical curve C_2 that lies on the tube of radius r centered at $\mathbf{x}(s)$.

The vector $\mathbf{B}$ is constant along C_1 and so $d\mathbf{B}/ds = 0$. Therefore (Exercise 5.4.5),

$$\frac{d\mathbf{V}(s)}{ds} = -\alpha \sin(\alpha s)\mathbf{N} + \cos(\alpha s)(-k\mathbf{T}) + \alpha \cos(\alpha s)\mathbf{B} \tag{5.9}$$

where k is the curvature of C_1 defined as the length of the vector $d\mathbf{T}/ds$ (incidently, we have implicitly assumed all along that r is smaller than the radius of curvature k so that the tube does not intersect itself).

We know that $\mathbf{B} = \mathbf{T} \times \mathbf{N}$ and it is easy to check that $\mathbf{T} \times \mathbf{B} = -\mathbf{N}$ and so $\mathbf{T} \times \mathbf{V} = \cos(\alpha s)\mathbf{B} - \sin(\alpha s)\mathbf{N}$. It now follows that

$$(\mathbf{T} \times \mathbf{V}) \cdot \frac{d\mathbf{V}}{ds} = \alpha \sin^2(\alpha s) + \alpha \cos^2(\alpha s) = \alpha$$

We are now ready to compute the twist using (5.8). This gives us

$$\text{Tw}(C_2, C_1) = \frac{1}{2\pi} \int_0^L \alpha \, ds = \frac{1}{2\pi} \int_0^L \frac{2\pi n}{L} \, ds = \frac{n}{L} \int_0^L ds = n$$

In this example, linking and twist are the same but this is not generally the case. If C_1 is allowed to wriggle about in space, a new quantity is introduced, call writhe, that measures how much wriggle there is. This can be done in a formal manner, as was done for the link and twist, but instead it is easier to invoke a fairly deep theorem that shows how all these terms are related, namely,

$$\text{Wr}(C_1) = \text{Lk}(C_2, C_1) - \text{Tw}(C_2, C_1) \tag{5.10}$$

The *writhe* of C_1, $\text{Wr}(C_1)$, is defined by this relation. It can be shown to be zero for curves confined to a plane, which explains why link and twist were the same in the example above.

Writhe is again dependent on how the curve is deformed and is not a topological invariant. Relation (5.10) is an intriguing mix of topological and geometrical notions that constrains how DNA can coil up.

Although it is possible to carry out some computations to evaluate writhe in specific cases, similar to but more demanding than the calculation already done (see the reference to the paper by White and Bauer [54]), it is more expedient to invoke our intuition at this juncture. Consider Figure 5.8. In the first panel (panel a) the link, twist, and writhe are all evidently zero, but in panel *b* the twist of C_2 about C_1 is still zero but the writhe of C_1 is not zero since the curve loops up and over itself out of the plane. In fact, the curves have a linking number of $+1$ and formula (5.10) shows that the writhe is also $+1$. Now deform C_1 so it doesn't loop out of the plane by undoing the loop of the ribbon up and over itself. The linking number remains unchanged under this deformation and we get the configuration of panel *c* in which the writhe is now zero (C_1 is planar) but the twist is $+1$ from formula (5.10). To verify the transition from *b* to *c* or back from *c* to *b*, simply take a piece of ribbon, give it a single twist and tape the ends together. Try it!

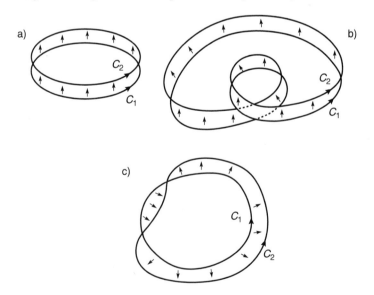

Figure 5.8. A ribbon model with different values of link, twist, and writhe.

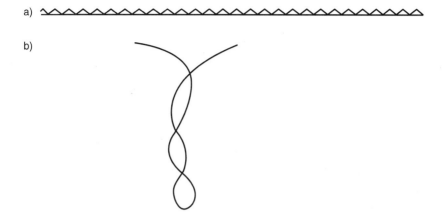

Figure 5.9. A string is held tautly by two ends (a) and then relaxed (b).

It is evident from these examples that as the curves are deformed in space, writhing and twisting are interchangeable quantities, even if the linking number remains unaltered. A more familiar example of this phenomenon is provided by grasping an ordinary string at its two ends, pulling it tight and then twisting it. The writhing is essentially zero, but the twist is considerable and can be felt in our fingers. Now bring the hands together. The twist seems to relax palpably, but the string loops (Figure 5.9) to form what is called a superhelix. The twist has now decreased, but the writhe is substantial. The same phenomenon is observed in a coiled telephone wire when it is stretched out. In this case, one goes from high writhe and low twist to just the opposite.

5.4. Exercises

5.4.1. Show that the divergence of the vector field $\mathbf{r}/\gamma^{3/2}$ is zero, where $\gamma = \mathbf{r} \cdot \mathbf{r}$ and $\mathbf{r} = (x - a, y - b, z - c)$.

5.4.2. Establish relation (5.4). (Hint: use Stoke's theorem of the multivariate calculus.)

5.4.3. Find the linking number of the curves in Figure 5.10a using Gauss's integral 5.7. It is convenient to deform C_1 by increasing its radius until it becomes a straight line through the center of C_2, as in Figure 5.10b. Choose C_2 to be a circle of radius one in the x, y plane and C_1 a line in the direction of the z axis. Then C_1 can be parametrized as $\mathbf{r}_1(\tau) = (0, 0, \tau)$ and C_2 by $\mathbf{r}_2(\theta) = (\cos \theta, \sin \theta, 0)$. The vector $\mathbf{r}$ is then $(\cos \theta, \sin \theta, -\tau)$.

5.4.4. First, some background to what is essentially the well-known Frenet-Serret apparatus of space curves. If a curve C is represented parametrically by $\mathbf{r}(s)$, with arc length s, then $\mathbf{T} = d\mathbf{r}/ds$ is a unit tangent to the curve at all points on C

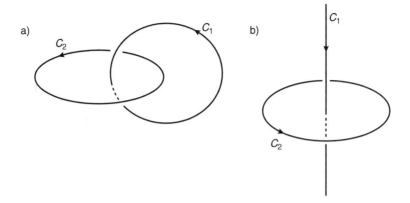

Figure 5.10. Two intersecting closed curves (a) and an infinite straight line intersecting a planar curve orthogonally (b).

defined by the parameter s. Since $\mathbf{T} \cdot \mathbf{T} = 1$ everywhere on C, the derivative of this scalar product is zero and so $\mathbf{T} \cdot d\mathbf{T}/ds = 0$. The length of $d\mathbf{T}/ds$ is called the curvature and therefore $d\mathbf{T}/ds = k\mathbf{N}$, where $\mathbf{N}$ is a unit normal to the curve (and orthogonal to $\mathbf{T}$). Let $\mathbf{B} = \mathbf{T} \times \mathbf{N}$. Since $\mathbf{T}$ and $\mathbf{N}$ have unit length so does $\mathbf{B}$, from the way a cross product of two vectors is defined.

Show that if C lies entirely in a plane then $\mathbf{T}$ and $\mathbf{N}$ also lie in the same plane, from which it follows that $\mathbf{B}$ is a constant unit length vector perpendicular to the plane. That is, $d\mathbf{B}/ds = 0$ along C. Now show that $d\mathbf{N}/ds = -k\mathbf{T}$ for planar curves.

5.4.5. Establish relation (5.9). (Hint: see the previous exercise.)

5.4.6. We know that $\text{Lk}(C_1, C_2) = \text{Lk}(C_2, C_1)$. Reversing the roles of the two curves in (5.10), we see that $\text{Tw}(C_1, C_2) - \text{Tw}(C_2, C_1) = \text{Wr}(C_2) - \text{Wr}(C_1)$. In the example carried out in Section 5.3 in which C_1 is a circle and C_2 is a helix wound about C_1, it is possible to compute the writhe of C_2 using the relation above between twist and writhe. It can be shown (in the paper by White and Bauer [54]) that $\text{Tw}(C_1, C_2)$ is approximately $nL/(L^2 + 4\pi^2 n^2 r^2)^{1/2}$. From this, establish that $\text{Wr}(C_2)$ is approximately $n(1 - (1 + 4\pi n r/L)^2)^{-1/2}$. This example also illustrates the fact that the twist of one curve about another is not necessarily the same if the two curves are interchanged.

5.5. Further Readings

Good background reading for DNA coiling is the *Scientific American* article [6], and the review in *Science* [21]. More technical details are provided in the paper [54] by White and Bauer. A proof of relation (5.10) is given in a paper by James White [53].

The proof of Gauss's formula (5.7) of Section 5.2 follows the one to be found in the book on multivariate calculus by Courant and John [24].

Measles and Blood Clots

6.1. Background

The setting is a crowded schoolroom in winter where one child after another comes down with a rash and fever, a case of measles. From time to time, epidemics of childhood diseases such as measles arise in cities around the country, most recently during the late 1980s. Susceptible individuals fall victim to a contagious disease when they encounter others who are already infected. There is a whole class of problems in which two or more species interact that can be formulated mathematically in quite similar ways. Grim warfare, territorial disputes, the hide-and-seek behavior of predators and their victims, and even chemical reactions, are all manifestations of species that affect each other because they are in the same place at the same time.

What the mathematical models of these processes have in common is that they record changes that take place over time and are described in terms of rates of change such as the rate of reproduction and the rate of becoming infected. But rates imply derivatives and so it is not surprising that the models are expressed as differential equations, a vast and sometimes difficult subject from which we extract a few essential ideas in the next two sections, enough so that we can gain some insight into an assortment of models. In addition to using analytical tools to reveal the behavior of solutions to the differential equations, it is also expedient to employ computer-generated graphics to display the numerically integrated solutions. This will be done throughout this chapter and the next.

The epidemic model itself is discussed in Section 6.4, and in Section 6.5 we obtain a brief glimpse of chaotic dynamics, especially as it relates to measles.

Another illustration of these ideas is the interplay of excitation and inhibition that is common in any number of living systems. A striking example is provided by blood coagulation in which the breaking of a blood vessel mobilizes a cascade of enzyme reactions that lead to the rapid formation of a clot to plug the lesion. Without this, one would bleed to death. However, an unchecked production of clots can also be fatal, and so the biochemical reactions include inhibitors that serve to brake the possibility of a runaway system. It is the delicate balance between these competing requirements that enables the blood-clotting system to work as well as it generally does. This topic is taken up in Section 6.6.

6.2. Equilibria and Stability

Because all the models in this chapter involve differential equations, we need to discuss some of the properties of simultaneous first-order ordinary differential equations in k dependent variables $x_i(t)$, $1 \leq i \leq k$. The x_i define the states of a dynamical system in terms of an independent variable t that is regarded as time:

$$
\begin{aligned}
x_1' &= f_1(x_1, \ldots, x_k) \\
x_k' &= f_k(x_1, \ldots, x_k)
\end{aligned}
\tag{6.1}
$$

The superscript prime denotes differentiation with respect to t (we suppress, for notational convenience, the explicit dependence on t in this and subsequent chapters), and the f_i are generally nonlinear functions of the k variables. A specific example of what we have in mind is the following system of two equations, a variant of which will be considered in the next chapter:

$$
\begin{aligned}
x_1' &= rx_1(1 - x_1) - x_1 x_2 \\
x_2' &= x_2(px_1 - c)
\end{aligned}
\tag{6.2}
$$

where r, p, and c are positive constants. The nonlinearity comes about from the fact that the right side of the equations include products of the variables. Other examples will follow later, but for the moment our discussion is necessarily more abstract.

Vectors are columns of real numbers, but for typographical convenience I represent them as transposes of row vectors. With this in mind, define

$$
\mathbf{x}(t) = (x_1(t), \ldots, x_k(t))^T, \qquad \mathbf{x}'(t) = (x_1'(t), \ldots, x_k'(t))^T,
$$

and

$$
\mathbf{f}(\mathbf{x}) = (f_1(\mathbf{x}), \ldots, f_k(\mathbf{x}))^T
$$

A compact version of (6.1) can now be written as

$$
\mathbf{x}' = \mathbf{f}(\mathbf{x})
\tag{6.3}
$$

The vector $\mathbf{x}(t)$ defines the state of the dynamical system at any time t and it traces out a solution curve in k-dimensional space, called a *trajectory or an orbit*, that describes how the system moves in time.

We assume that the equations are an adequate representation of an observable process, and so on purely intuitive grounds, it is reasonable to expect that the equations have solutions that mimic the actual time evolution of the process itself. Specifically we assume that if the state is prescribed to have a value $\mathbf{x}_0$ at some initial time $t = 0$, then there is a unique vector valued function $\mathbf{x}(t)$ defined for all times t that passes through $\mathbf{x}_0$ at $t = 0$. Precise conditions on the vector function $\mathbf{f}$ can be given that guarantee that a unique solution with these properties does indeed exist but we forgo the details because we already believe that the equations are well posed in terms of the processes that they describe. Any good book on the elementary theory of differential equations can be used to fill in the mathematical details that are sidestepped here; Chapters 5 and 6 of Jeffrey's book [31] can be recommended. The uniqueness requirement is especially significant because it tells us that the time evolution of a dynamical system along individual trajectories follows distinct paths; if two trajectories were to cross each other, then their point of contact is an initial state from which two separate solutions emerge, a violation of uniqueness.

Equations (6.1) or (6.3) are a first-order system because they involve only the variables x_i and their first derivatives. A single kth-order scalar equation that has derivatives up to order k can always be reduced to a set of k first-order equations. For example, the second-order equation

$$y'' + \sin y = 0$$

becomes

$$x_1' = x_2$$
$$x_2' = -\sin x_1$$

when we let $x_1 = y$ and $x_2 = y'$.

Some solutions remain constant in time. These states of rest are called *equilibrium solutions* and are defined by positions for which the time derivative $\mathbf{x}'$ is zero. At an equilibrium no motion ensues. In Equations (6.2), for example, the equilibrium states are the vectors $(0, 0)^T$, $(1, 0)^T$, and $(c/p, r(1 - c/p))^T$.

Our concern is with knowing what happens to a trajectory as time evolves. An equilibrium state $\mathbf{q}$ is called an *attractor*, and is said to be *asymptotically stable*, if all trajectories that begin in some sufficiently small neighborhood of $\mathbf{q}$ approach this equilibrium arbitrarily close as t increases. More specifically, there is a set Ω of initial states $\mathbf{x}_0$ such that the solution through $\mathbf{x}_0$ will lie in an arbitrarily small neighborhood of q when t is large enough. The largest set Ω for which this is true is called the *basin of attraction* of $\mathbf{q}$. A state $\mathbf{q}$ may fail to be a point attractor if trajectories that begin close enough to $\mathbf{q}$ remain confined to some bounded neighborhood of $\mathbf{q}$ for all time but do not actually tend to $\mathbf{q}$. In this case, we say that $\mathbf{q}$ is stable, but not asymptotically stable, whereas if no such bounded neighborhood exists then $\mathbf{q}$ is called *unstable*. An unstable equilibrium repels at least one trajectory in its neighborhood.

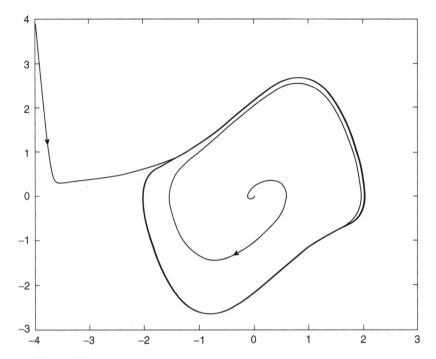

Figure 6.1. A closed orbit of the Van der Pol equation as a periodic attractor.

There are attractors other than equilibria. Trajectories that begin in some set Ω of initial states may, for example, tend toward a closed orbit that represents a periodic solution of the differential equations. An illustration of this is provided by the *Van der Pol equation* $y'' = -y + y'(1 - y^2)$, which becomes a first-order system

$$x_1' = x_2$$
$$x_2' = -x_1 + x_2(1 - x_1^2)$$

when we let $x_1 = y$ and $x_2 = y'$.

The solutions in x_1, x_2 space are plotted in Figure 6.1. There is a single equilibrium state at the origin that is unstable because it repels all trajectories in its neighborhood. These trajectories spiral outward to the cyclic trajectory, whereas all solutions that begin outside the closed orbit eventually wind down toward it.

There are other, more complicated examples of attractors to be discussed later, but for now our attention will be confined to asymptotically stable equilibria.

Our task now is to identify point attractors of specific equation systems. The simplest case is that of a single nonlinear equation in which x is a scalar function of t (namely, when $k = 1$). Although it is sometimes possible to solve the equations

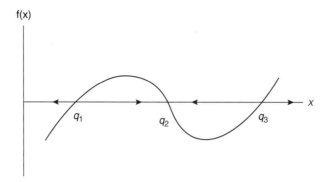

Figure 6.2. Stable and unstable equilibria of a scalar equation.

explicitly there are many interesting cases in which there is a simpler, more direct approach to the question of what happens to trajectories as t increases. One begins by finding the equilibria of $x' = f(x)$, which is the algebraic problem of finding the roots of the equation $f(x) = 0$. If q is one such root, consider the sign of $f(x)$ in a neighborhood of q. If $f(x)$, and therefore the derivative of $x(t)$, is positive, this tells us that x is increasing, whereas a negative sign indicates that $x(t)$ is decreasing. In Figure 6.2 we consider a typical $f(x)$ that crosses the x axis at three points q_1, q_2, and q_3. These are the equilibria of $x' = f(x)$ and by looking at the sign of f we immediately verify that q_1 and q_3 are unstable and q_2 is an attractor. The arrows indicate the direction of the one-dimensional trajectory $x(t)$.

To make our discussion more concrete we now derive an equation that models a growth and decay process of considerable interest. An organism is assumed to grow at a constant per capita rate r. This means that the net difference in the rate of births and deaths, per unit of population, is constant. Thus, if three cells per thousand divide every minute and one per thousand dies, the per capita rate r is .002. Let $x(t)$ denote the total population at time t, with an initial count of $x(0) = x_0$. Taking r to be positive, the rate of increase in population is simply $x'(t) = rx(t)$. This differential equation has a correspondingly simple solution, as is readily verified: $x(t) = x_0 e^{rt}$. It predicts that the population grows exponentially from its initial value. Although this is roughly true for a number of species for relatively short periods of their history, including humans, it grossly exaggerates what happens over longer time spans. A better model is obtained by recognizing that as a population grows the eventual overcrowding causes the birth rate to decline and the death rate to increase because of disease and aggressive competition for dwindling resources. This suggests that a constant r should be replaced by a per capita rate that decreases as population increases. A plausible candidate is obtained by replacing r with $r(1 - x/K)$, where K represents the theoretically maximum population that can be sustained in a given habitat. Suppose now that the population is subject to exploitation, such as a fishery, or to predation, such as algae in the ocean that are grazed on by small crustaceans, or even emigration from a region in the case of humans. Let $E > 0$ denote the per capita rate of removal of the species from the region in which it finds itself. Then the equation for the rate of growth of the population now becomes

$$x' = rx\left(1 - \frac{x}{K}\right) - Ex \tag{6.4}$$

The first term on the right is the net growth of the species due to its own internal dynamics of birth and death, whereas the second term captures the external effect of movement across the region's boundary. Incidentally, if E is taken to be negative this represents a net inflow into the region, such as immigration for humans or the restocking of a fishery, or a massive migration of algae into a part of the ocean due to a sudden shift in the current.

Equation (6.4) can be rewritten as

$$x' = (r - E)x\left(1 - \frac{x}{K\left(1 - \frac{E}{r}\right)}\right) \tag{6.5}$$

and there are evidently two equilibria, at $x = 0$ and $x = K(1 - E/r)$. If one graphs the right side of the equation, namely, $f(x)$, it is seen that it is positive between zero and $K(1 - E/r)$ and negative otherwise. From this we deduce, as before, that zero is unstable while $K(1 - E/r)$ is an attractor. This means that the population will inevitably increase from small values to a maximum value that approaches K as E is reduced to zero. Is this reasonable? Well, in some situations, such as bacteria confined to a beaker supplied with nutrients, this seems to be what happens and it is an acceptable approximation for human populations as well. However, the model is based on several questionable assumptions that suggest caution in its use. We return to this in the next section but for now our goal is to examine the mathematical side of this and other nonlinear equations.

An alternative approach to deciding whether an equilibrium is attracting is based on the idea of approximating the nonlinear equation $x' = f(x)$ by a linear equation that is readily solved. If q is an equilibrium state, Taylor's theorem in the calculus allows us to write $f(x)$ as $f(q) + f'(q)(x - q) +$ higher order terms in $x - q$ (we assume that f is a twice differentiable function). For x sufficiently close to q the higher order terms are negligible and so if one ignores the terms in $(x - q)^2$ and higher, the original equation can be approximated by the linear equation $u' = au$, where $a = f'(q)$ and $u = x - q$ (recall that $f(q) = 0$). This provides a *linearization* of $x' = f(x)$ about the equilibrium q and it is reasonable to conjecture that in a neighborhood of q the solution of the nonlinear equation shadows the linearized solution. Equation $u' = au$ has the solution $u(t) = u_0 e^{at}$, where u_0 is the initial value of u, namely, $x(0) - q$. If a is negative it is apparent that $u(t)$ tends to zero, its equilibrium solution, as t goes to infinity. When a is positive, the solution grows without bound and the origin is unstable. Now a movement of u toward zero is tantamount to x going to q, and so it would seem that q is an attractor for the nonlinear equation if zero is an attractor for its linearized version. This leads to the next result, which verifies our conjecture.

Lemma 6.1   *Let q be an equilibrium of the scalar equation $x' = f(x)$, where f is twice differentiable, and let $a = f'(q)$. Then q is an attractor if $a < 0$ and it is unstable if $a > 0$.*

Proof: Let $u = x - q$ to write the equation $x' = f(x)$ as $u' = au + g(u)$, where $g(u)$ represents terms in u of order 2 and higher. The function g satisfies $g(0) = 0$ and $g'(0) = 0$. Moreover g' is continuous and, therefore, for any given ε, $|g'(u)| < \varepsilon$ for all $|u| < \delta$, where δ is suitably small. But

$$g(u) = \int_0^u g'(s)\,ds$$

and so $|g(u)| < \varepsilon|u|$ when $|u| < \delta$. Suppose that a is negative. Then $a < -\varepsilon$ for ε small enough. This means that $u' = au + g(u) < -\varepsilon u + \varepsilon u = 0$ provided that $0 < u < \delta$. In the same way, u' is positive if $-\delta < u < 0$. In either case it follows that if $|u| < \delta$ initially, then $u(t)$ goes to zero as t increases. When a is positive a similar argument shows that the origin is unstable. The result follows by noting that $x = u + q$.   □

Thus we have resolved the issue of stability or instability for the case of scalar equations. Of course one can sometimes, though not always, explicitly solve the equation and thereby obtain the long term behavior of the solutions directly, as in the case of Equation (6.5) (Exercise 6.7.1), but as soon as we go beyond the case of single equations to systems, the likelihood of an explicit solution is sharply reduced and one is constrained to use surrogate methods. We illustrate one such technique for pairs of equations ($k = 2$) that turns out to be a two-dimensional analogue of what we did above in the scalar case. An additional method will be discussed in Section 6.3.

To be specific, we consider the situation of coupled equations that model the competition of two species that cohabit a given region with populations of x_1 and x_2, respectively. Each species can thrive in isolation from the other and separately follow a growth law of the form of (6.4) in which $E = 0$. In the biological literature this is known as a *logistic equation*. However, when the two species are forced to compete with each other for food and other limited resources, each inhibits the other and diminishes the ability of either of them to fully prosper. This a form of mutual exploitation that we model by terms that are proportional to the product of their populations since this measures the likelihood of an encounter between them:

$$x_1' = rx_1\left(1 - \frac{x_1}{K}\right) - ax_1x_2$$

$$x_2' = sx_2\left(1 - \frac{x_2}{L}\right) - bx_1x_2$$

(6.6)

All constants are positive in which s is the intrinsic per capita growth rate of the second species and L is the maximum population level this species can theoretically sustain within the given region. The coefficients a and b measure the competitive advantage of one species over another. As a increases, for example, this indicates that second species becomes more effective in outwitting, subduing, or otherwise inhibiting its competitor. Of course, this model is only a caricature of what really takes place between two or more species and its utility lies in the fact that it is a widely adopted paradigm of the raw struggle for dominance in nature. We will have occasion to treat a generalization of these equations in a later chapter that explicitly includes the effect of movement in space, something that is only implicit in the present version in which the spatial background is "frozen."

Equations (6.6) have several equilibria. The algebraic equations obtained by letting x'_1 and x'_2 be zero separately are called *nullclines* and the intersection of these nullclines are the points in the plane where the equilibrium states occur since both of the right-hand sides are simultaneously zero there. An easy computation shows that there are four possibilities for equilibria in the positive quadrant (any solution with either of the x_i negative makes no sense physically and is disregarded). To begin with there is the uninteresting case in which both species are zero, then two possibilities in which one or the other species, but not both, is zero, and the case in which the nullclines intersect at a positive value for both species. We pick a specific situation and consider parameters for which $r/L < a$ and $s/K < b$. This forces the nullclines to intersect in the positive quadrant where neither species is zero. This is shown in Figure 6.3. Other possibilities are considered in Exercise 6.7.2.

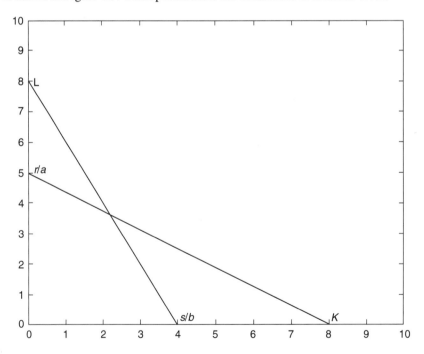

Figure 6.3. Nullclines for the competition model, Equations (6.6).

We now return to the idea of looking at the sign of the derivatives, a ploy that worked so well for us in the case of a single scalar equation.

Factor out the positive equilibrium values of x_1 and x_2 from the right-hand sides of (6.6). The nullclines are the straight lines exhibited in Figure 6.3 and a little reflection shows that if x_1 has a value that lies to the right of the line defined by $x_2' = 0$, then the derivative of x_2 is negative, whereas if its value lies to the left of this line, the derivative is positive. Similarly, when x_2 is a value that lies above the line defined by nullcline $x_1' = 0$ then the derivative of x_1 is negative and it changes sign when x_2 is below the line. The signs of the derivatives tell us, of course, whether the variables x_1 and x_2 are increasing or decreasing and this gives us a clue to the direction of flow of the solution curves, namely, the trajectories. Note that these curves must cross the nullclines either vertically or horizontally. For example, if $x_1' = 0$ it means that there is no change in the x_1 variable and so the trajectory must be vertical as it crosses the nullcline defined by this zero derivative.

Putting all this information together enables one to make a crude sketch of the flow of the trajectories, a useful task that provides considerable insight into the dynamics of the model equations. To illustrate the method consider the region that lies above and to the right of the two lines in Figure 6.3. Both derivatives are negative here, and so the flow is simultaneously downward and to the left. Because all trajectories are smooth curves (the functions f_i of the differential equations are assumed to be at least differentiable), the way in which they bend is determined by the whether they must cross the nullclines horizontally or vertically. Some trajectories are forced to bend in one direction while others bend in the opposite way. This is illustrated in Figure 6.4, where typical computer-generated trajectories appear from all portions of the quadrant. The orbits were obtained by a numerical solution of the equations. Note that the horizontal and vertical axes represent valid solutions to Equations (6.6) because these correspond to setting either x_1' or x_2' to zero. It follows that no trajectory that begins in the positive quadrant can ever leave this domain, because otherwise it would have to cross one of these axes and, as we observed earlier, this would violate the property of uniqueness of solutions. Therefore the trajectories shown in Figure 6.4 approach, but never actually reach, the equilibria.

Looking at Figure 6.4 one sees that virtually all solutions tend to either of two attractors in which one species or the other is zero, depending on their initial state. Apparently the species intimidate each other sufficiently that one of them is eliminated, a phenomenon called *competitive exclusion*. The only exception to this occurs along the pair of curves that tend to the unstable equilibrium in which both species coexist. These curves define what is called a *separatrix* since they separate all other trajectories into two mutually exclusive classes, those that tend to $(0, L)^T$ or to $(K, 0)^T$. This precariously delicate coexistence depends on choosing the initial values of the variables just right, an improbable feat. All states, other than those on the separatrix or the origin, belong to the basin of attraction of one of these two attractors. The origin is an unstable equilibrium.

In conclusion we see that fierce competitiveness between two species inevitably leads to the extinction of one of them. In effect, they cannot occupy the same

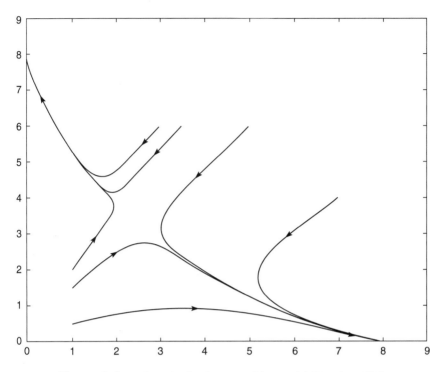

Figure 6.4. Trajectories for the competition model, Equations (6.6).

ecological niche. On the other hand, a different choice of the competition constants *a* and *b* in the model can lead to coexistence, implying that each species occupies distinct or, at most, partially overlapping niches. Competition, in this case, is mild (Exercise 6.7.2).

Another, more analytical approach to the same problem mimics the linearization procedure used earlier for scalar equations and is discussed in the next section.

6.3. Linearization

The intuitively satisfying nullcline approach outlined is a sufficiently good tool for several of the models that we deal with in this chapter and elsewhere in the book. However, it is effectively limited to problems in the plane, where pictures can be drawn and, even here, it sometimes cannot unambiguously resolve what the trajectory portrait actually looks like. In these cases, there is another, more analytical approach to the same problem that mimics the linearization procedure used earlier for scalar equations. For simplicity of notation we restrict ourselves to planar problems but our conclusions remain the same, word for word, when **x** defines trajectories in R^k instead of R^2.

As before, let $\mathbf{q}$ be an equilibrium vector and define $\mathbf{u} = (u_1, \ u_2)^T$ to be $\mathbf{x} - \mathbf{q}$. The equation system is

$$
\begin{aligned}
x_1' &= f_1(x_1, x_2) \\
x_2' &= f_2(x_1, x_2)
\end{aligned}
\tag{6.7}
$$

with $\mathbf{q} = (q_1, q_2)$. Taylor's theorem for functions of two variables enables us to expand f_i about q for each $i = 1, 2$ if these functions possess twice-differentiable partial derivates in the variables x_1 and x_2, as is generically assumed. Because $\mathbf{f}(\mathbf{q}) = 0$, we obtain

$$
f_i(x_1, x_2) = \frac{\partial f_i}{\partial x_1}(q_1, q_2)u_1 + \frac{\partial f_i}{\partial x_2}(q_1, q_2)u_2 + g_i(u_1, u_2)
$$

where the g_i are higher order terms in u_1 and u_2, meaning by this that the ratio of $g_i(u_1, u_2)$ to the Euclidean norm of u can be made less than any preassigned ε, as soon as the norm of u is smaller than some suitably chosen δ that depends on ε. If these terms in g_i are ignored, the two equations for f_i can be combined into a single matrix statement

$$
\mathbf{u}' = \mathbf{Au}
\tag{6.8}
$$

where $\mathbf{A}$ is the Jacobian matrix of f, evaluated at $\mathbf{q}$:

$$
\mathbf{A} = \begin{pmatrix} \dfrac{\partial f_1}{\partial x_1} & \dfrac{\partial f_1}{\partial x_2} \\[2mm] \dfrac{\partial f_2}{\partial x_1} & \dfrac{\partial f_2}{\partial x_2} \end{pmatrix}
$$

Equation 6.8 is the linearization of nonlinear system (6.7) and it has a well-known solution in terms of the eigenvalues and eigenvectors of the matrix $\mathbf{A}$ (see the book by Jeffrey [31] for a clear treatment of the theory of systems of linear differential equations). Assuming the invertibility of $\mathbf{A}$ it is clear that $\mathbf{u} = \mathbf{0}$ is the only equilibrium for $\mathbf{u}' = \mathbf{Au}$.

We do not need to explicitly solve linear systems, but are more concerned with the following results which are stated without proof:

Lemma 6.2 *Let $\mathbf{q}$ be an equilibrium for the planar nonlinear system $\mathbf{x}' = \mathbf{f}(\mathbf{x})$, and let $\mathbf{A}$ be the Jacobian matrix of $\mathbf{f}$ evaluated at $\mathbf{q}$. Then $\mathbf{q}$ is an attractor if the real parts of the (possibly complex) eigenvalues of $\mathbf{A}$ are both negative, and $\mathbf{q}$ is unstable when this is not true.*

An explicit and useful condition to insure that the eigenvalues satisfy Lemma 6.2 is given in the next result.

Lemma 6.3   *The equilibrium solution* **q** *is an attractor if the trace of* **A** *is negative and the determinant of* **A** *is positive. When the determinant of* **A** *is negative then, regardless of the sign of the trace, the eigenvalues are real and of opposite sign and* **q** *is unstable.*

Lemma 6.4   *When the eigenvalues are real and of opposite sign there is a trajectory to the nonlinear system that, in the vicinity of the unstable equilibrium* **q***, appears as a smooth curve tangent to the eigenvector corresponding to the negative eigenvalue of* **A***. Along this curve, called the separatrix, the solution tends to* **q***. Another smooth curve is tangent to the eigenvector corresponding to the positive eigenvalue of* **A** *and, along this curve, there is a solution that moves away from* **q***.*

Although Lemmas 6.2 and 6.3 are fairly standard facts in the study of differential equations, the last lemma is a planar version of a more sophisticated assertion known as the stable and unstable manifold theorem. By carefully reformulating the conditions, one can restate these results as theorems in R^k for any k.

We can illustrate the applicability of these theoretical results by returning to the competitive model Equations (6.6). Suppose that $r/L < a$ and $s/K < b$, in which case the equilibria are both positive. The Jacobian matrix is then readily evaluated at $\mathbf{x} = \mathbf{q}$ to be

$$\begin{pmatrix} -rx_1/K & -ax_1 \\ -bx_2 & -sx_2/L \end{pmatrix}$$

It is straightforward to see that the trace and determinant of this matrix are both negative and this implies an unstable equilibrium by Lemma 6.3. To complete our vindication of the nullcline approach carried out previously, it suffices to consider one of the two nontrivial equilibria located along a coordinate axis. When $x_1 = 0$ and $x_2 = L$, for instance, the Jacobian matrix becomes

$$\begin{pmatrix} r - aL & 0 \\ -bL & -s \end{pmatrix}$$

and, since it is still true that $r < aL$ and $s < bK$, the determinant is now positive although the trace continues to remain negative. By Lemma 6.3 the equilibrium is an attractor and this is also seen in Figure 6.4.

6.4. Measles Epidemics

Our intention is to model childhood epidemics such as measles, chickenpox, and mumps to track the dynamics of outbreaks of these contagious diseases.

We begin by dividing up the population of a given region into four categories: susceptibles are those individuals who are able to contract the communicable disease, exposed are infected people who are not yet able to transmit the disease to others, infectives are those who are capable of spreading the disease, and recovered.

With childhood illnesses it is largely true that a recovered person becomes permanently immune, and so the last category includes individuals who are immune to begin with or who succumb to the disease. Denote the fractions of susceptible, exposed, infective, and recovered in the population by S, E, I, and R, respectively. We assume that all newborns are susceptibles at birth, and that the population size remains constant by balancing births, on one hand, to deaths and emigrations, on the other. The diseases are endemic in the sense of recurring from year to year, with varying severity. As susceptibles become ill and, ultimately, immune or dead, their supply is replenished by newborns and so the model must explicitly include birth rates. Later we consider what happens in virulent diseases whose contagion is so rapid that it runs its course within a time span so short that the addition of fresh susceptibles can be disregarded.

Denote by b the disease contact rate, which is the average number of effective disease transmissions per unit time per infective. This requires a word of explanation. The disease is transmitted in proportion to the number of possible encounters between infectives and susceptibles, which means that b is the average rate at which an infective comes into contact with another person and so bI is the total contact rate. This has to be multiplied by the likelihood that the person contacted is in fact a noninfective and this equals S, the fraction of susceptibles in the population. It is assumed that b increases with the total population size since the number of contacts is greater in dense urban areas than in sparse rural aggregations. This situation may not prevail, however, when the number of susceptibles is in large excess over the quantity of infectives. Increasing S in this case has little or no effect on the number of contacts made by an infective and the limiting factor now for disease transmission is only the size of I. An example of this is discussed in Exercise 6.7.6.

Let r be the average birth rate ($1/r$ is therefore the average life expectancy). Then

$$S' = r - rS - bSI \tag{6.9}$$

models the rate of change of S. The second term on the right expresses the fact that, because total population is constant, susceptibles are removed by death and emigration at the same rate r that they enter the population.

The class of exposed individuals increases over time at a rate equal to the rate at which susceptibles become infected, namely, bSI. They disappear from the population at a rate rE, due to death and emigration, and at a rate aE, where a is the average number of exposed people that become contagious per unit time. The reciprocal $1/a$ is, therefore, the mean latency period during which the disease incubates prior to becoming infectious. Putting this together gives the equation

$$E' = bSI - rE - aE \tag{6.10}$$

The infectives are now seen to increase at a rate aE and to be removed at rates rI (for the same reason as before) and cI, where c is the average number of infected people that recover per unit time; the reciprocal $1/c$ is then the mean infectious period. Thus

$$I' = aE - rI - cI \tag{6.11}$$

Finally, the recovered class evidently increases at a rate cI and disappears at rate rR:

$$R' = cI - rR \tag{6.12}$$

The four coupled equations describe the epidemic model but are too formidable to handle using the tools at our disposal. To get some idea of the behavior of the solution to this system we simplify a bit and ignore the class E by assuming that the latency period is zero. This means that susceptibles are immediately infected and so Equation (6.11) for I needs to be replaced by

$$I' = bSI - rI - cI$$

Three equations remain:

$$S' = r - rS - bSI$$
$$I' = bSI - rI - cI \tag{6.13}$$
$$R' = cI - rR$$

The first two equations are independent of R and so we can simplify further by retaining only these two. Because $S + I + R = 1$, the value of R is obtained from a knowledge of S and I. Thus

$$S' = r(1 - S) - bSI$$
$$I' = I(bS - (r + c)) \tag{6.14}$$

Using the heuristic nullcline approach described in Section 6.2 (with details worked out in Exercise 6.7.5) we surmise that there are oscillatory solutions about the equilibrium state defined by $S = (r + c)/b$, $I = r(b - r - c)/b(r + c)$. The quantity $(r + c)/b$ is taken to be less than one to guarantee that I is positive at the equilibrium. Therefore, if the ratio of the recovery rate c to the contact rate b is small enough, the epidemic can be sustained. This implies that intervention on the part of public health officials can have an impact on the severity or even the likelihood of an epidemic. For example, inoculation of school children reduces the number of susceptibles, while quick isolation of infectives increases the removal rate c.

The trajectories for a typical case were generated by computer and are shown in Figure 6.5. We observe that the oscillations wind down to a point attractor as the number of infectives waxes and wanes. A numerical integration of the full set of Equations (6.9) to (6.12) would reveal a similar pattern of damped

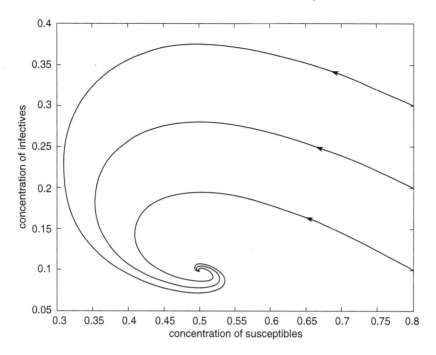

Figure 6.5. Trajectories for the epidemic model of Equations (6.14).

oscillations, which appears to indicate a cyclic recurrence of outbreaks with diminishing severity.

The Jacobian for Equations (6.14) at the equilibrium values of S and I is calculated quite simply and is given by

$$\begin{pmatrix} -r - bI & -bS \\ bI & 0 \end{pmatrix}$$

From this we see that the equilibrium is indeed an attractor, according to Lemma 6.3. Also, the eigenvalues are complex, which indicates oscillatory solutions (as shown in the book by Jeffrey [31]). This is what Figure 6.5 shows.

The actual data on measles show a different pattern in which the disease recurs with levels of severity that fluctuate erratically from year to year. In New York City, for example, measles outbreaks from 1928 to 1963 (when large-scale vaccination of schoolchildren began), the monthly data on measles cases exhibit irregular peaks and troughs that appear to have two-year cycles of highs and lows (Figure 6.6).

There are several possible reasons for the discrepancy between what the model predicts and what one observes. The most compelling, perhaps, is that the parameter b, the rate at which the disease is transmitted, need not be constant and can, in fact, vary widely. The disease pathogen appears to be more virulent during winter months at a time when children are confined together in school; contact is not

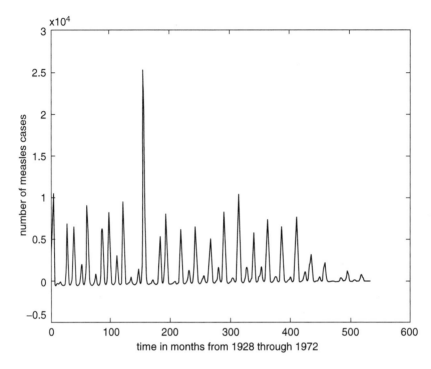

Figure 6.6. Actual measles cases in New York City averaged monthly from 1928 through 1972.

as close and is less frequent during the summer months. With this in mind the parameter b can be replaced in the model by a periodic function having a period of 1 year, with a low value in July and a high one in February:

$$b(t) = b_0(1 + b_1 \cos 2\pi t)$$

where b_0 is the average contact rate and b_1, a number between zero and one, measures the magnitude of the seasonal effect. The time $t = 0$ represents the beginning of February. With $b(t)$ substituted into the equations in place of a constant b, the original model can be integrated numerically to give solutions that indeed mimic the data seen in New York City and elsewhere for actual measles outbreaks. In certain parameter regimes, in fact, the periodically forced solutions are indistinguishable from random fluctuations, a phenomenon that some authors describe as a manifestation of chaotic dynamics in which, loosely speaking, the trajectories settle down to a complicated attractor in state space that is neither a point nor a cycle. More will be said about chaotic dynamics in Section 6.5.

It is important to note that all the parameters in the model, namely, r, b, b_o, and c, are statistical averages from actual data and so the equilibrium values of S and I are also averages. However, stochastic fluctuations in the actual values of these variables, such as those treated in Exercises 3.6.9 and 3.6.10 in Chapter 3, can be significant when the actual values of the variables are quite small. If the number of infectives is low, for instance, a chance fluctuation could reduce it to zero. In this case, the epidemic is extinguished, a condition known as fade-out. This can also be seen during those periods in which a contagious disease is waning because the number of susceptibles has decreased sufficiently that the rate of arrival of new susceptibles is not fast enough to compensate for the die-off or immunity rate of the infectives. Random events can exacerbate this situation by the sudden removal of people during one of the troughs in a wax and wane cycle, with the consequence of quenching the transmission of disease. For these reasons it is generally assumed, for measles anyway, that the population of susceptibles is substantial. It is known, in fact, that measles cannot remain endemic in communities with much less than a million people.

When an epidemic comes and goes quickly in one cycle the influx of new susceptibles can be neglected in the model Equations (6.14) by choosing r to be zero. Looking at the sign of the derivative in the equation for I (when r is zero) shows that the number of infectives is on the rise only when the initial fraction of susceptibles is greater than c/b. This is called a *threshold condition* because above it there is an initial increase in the number of contagious individuals, whereas below it the epidemic does not ignite. This is quite similar to what happens in the autocatalytic model of a biochemical reaction that will be discussed in the subsequent section (see also Exercise 6.7.6.)

6.5. Chaotic Dynamics or Randomness?

Let's return now to the measles model of the preceding section, which was found to be wanting because it failed to mimic actual measles events from year to year in a city like New York. It was suggested there that more realistic results could be obtained by making the contact rate b into a periodic function to simulate seasonal effects.

Figure 6.7a shows the fraction I of infective cases over a 30-year period as obtained from an integration of the periodically forced Equations (6.9) through (6.12). The parameters in the model are estimated from census data and medical records as $r = .02$ per year, $b_0 = 1800$ per year, $a = 35.58$ per year, $c = 100$ per year, and $b_1 = .285$. The severity of the outbreaks varies erratically from year to year and resembles that found in the actual New York City data from 1928 to 1963, during which the epidemic peaked at about 25,000 cases a year (Figure 6.6). With a total population of about 6 million, this corresponds to a maximum fraction of infectives of roughly .004, as shown in Figure 6.7a.

The solutions of the model equations have evidently settled down to an attractor that is clearly recurrent from year to year, but the magnitude of the fluctuations is somewhat unpredictable. This is either a manifestation of a cyclic attractor having a very long period of motion or it may represent what is known as a *chaotic attractor*. There are two important signatures of a chaotic attractor or, more simply, *chaos*.

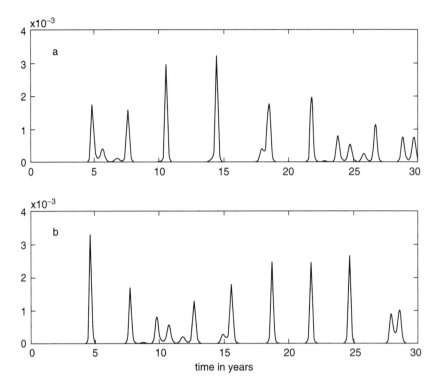

Figure 6.7. The fraction of infectives over a 30-year period from the periodically forced model Equations (6.9) through (6.12), with $b_1 = .285$. The only difference between panels a and b is that the initial values of susceptibles S differs by .00001, with all other model parameters remaining the same. The resulting orbits demonstrate a sensitivity to initial conditions, a signature of chaotic dynamics.

One of these is an extreme sensitivity to initial conditions, meaning that small changes in initial values of any of the variables produces solutions that rapidly diverge from each other over time. This evidently cannot happen on a periodic attractor. The other attribute of chaos is that orbits will move arbitrarily close to every point on the attractor infinitely often. The time lapse between successive returns is erratic, however. This recurrent behavior differs from that of a periodic attractor because the orbits weave a complex tangle that never cross themselves (if they did intersect, the uniqueness principle would force the trajectory to be a closed path on which motion is periodic).

To illustrate the idea of sensitivity, let us change the initial value of S by only .0001 from the value used to obtain Figure 6.7a. All other initial values, for I, E, and R remain the same. The result is shown in Figure 6.7b, where we see that the solution remains roughly the same for a while to the one shown in the figure above it, but soon it changes dramatically so that by the time 30 years have elapsed the value of I is evolving along a completely different part of the attractor. Moreover, although there are annual measles outbreaks, the magnitude of the fluctuations in the number of infectives is unpredictable and the time between nearly recurrent events appears to be essentially random. This shows that we can never hope to replicate exactly the actual dynamics of a chaotic process, and the solutions only illustrate what the typical behavior is like.

These numerical results are still somewhat unsatisfactory, however, since the actual New York City data from 1945 onward exhibit alternating highs and lows in the number of reported cases of measles with a severe outbreak in 1 year followed by a mild incidence of disease in the subsequent year. The problem is that the parameter b_1 that measures the effect of seasonality is not known exactly and even a slight change in its value can have a notable effect in the computational results. For example, a change to $b_1 = .275$ gives alternating biannual peaks in the number of cases as shown in Figure 6.8, which is more consistent with the data from 1945 onward but, on the other hand, doesn't reach the reported levels of severity that were visable in Figure 6.7, where $b_1 = .285$.

Although the model is a plausible caricature of the actual dynamics of a recurrent epidemic it is at least arguable whether an alternative formulation would not be more convincing. One can allow the parameters to vary at random, for example,

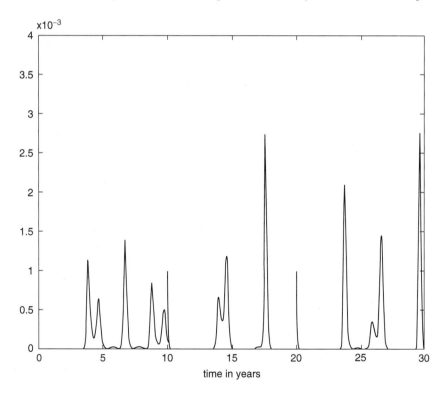

Figure 6.8. Fraction of infectives over a 30-year period from the periodically forced model Equations (6.9) through (6.12), with $b_1 = .275$.

to represent chance variations, as in the wayward movement of infectives into and out of the region. Chance events that disturb the rhythm of periodic dynamics sometimes can give the appearance of chaos and some systems modeled by three or more differential equations have attractors that are not quite periodic but nevertheless an appearance of recurrence is maintained in the sense that the orbit returns to nearby points infinitely often, weaving a complex "stretch and fold" pattern in some confined region of space.

Attractors for certain biological systems seem to behave this way either because the motion represented by a cycle loops around for a very long time before returning to its starting point, or because the limiting behavior consists of fluctuations that appear random. Indeed, the populations of many species exhibit erratic fluctuations and although some of the recurrence that one observes may partially be due to an internal dynamic that determines the interactions with other species and its environment, there may also be a component of chance. As we saw in the case of measles, it is difficult to distinguish in an unambiguous manner whether one is dealing with deterministic chaos or noise contamination that masks the biological interactions. Although the footprints of chaos can be partially discerned in some data sets, a clear marker is still lacking.

Another example of this ambiguity, besides that of measles, is connected to the tiny marine diatoms that populate many water bodies. Figure 6.9 shows the weekly

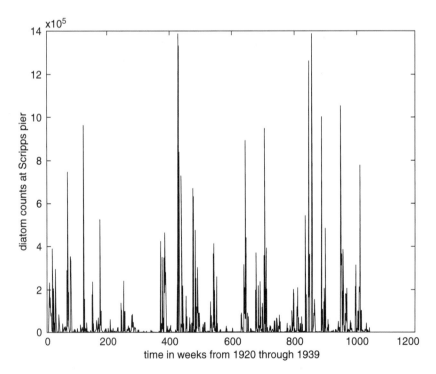

Figure 6.9. Weekly averaged diatom counts at Scripps pier from 1920 through 1939.

average diatom count over a span of 20 years as measured near the pier of the Scripps Institute in La Jolla, California. There are several known and conjectured influences on the rise and fall of the diatom counts including changes in temperature and salinity, the upwelling of nutrients from deeper waters, and the effects of predation. Nonetheless the haphazard fluctuations recorded in Figure 6.9 may be a manifestation of randomness as much as any orderly underlying process.

The allure of chaos is that it enables us to think of complex fluctuations as a deterministic process described by differential equations, even when the observed motion appears to be the result of random noise superimposed on some orderly process.

6.6. Blood Clotting

Running up the stairs a boy trips and scrapes his knee. Blood begins to ooze from the lesion but not for long. The broken blood vessels have released something called *tissue factor* (TF) that reacts with a protein in the bloodstream to form a complex that we call Z_1. This begins a chain of reactions that ultimately leads to the formation of fibrin, a protein that is polymerized into the gelatinous clump that we know as a blood clot.

The initiation of this blood-clotting cascade of biochemical interactions first requires that Z_1 be converted into an active form before it can be useful. This occurs by means of an enzyme E_2 that cleaves Z_1 into its activated form. The letter Z stands for zymogen, an enzyme precursor, and the activated form of Z_1 is another enzyme that we call E_1. This enzyme then interacts with yet another protein Z_2 in the bloodstream, also known as factor X, to form the very same enzyme E_2 that initiated these reactions in the first place and, by so doing, it stimulates its own production. E_2 then goes on to interact with other blood proteins besides Z_1 until the full set of reactions has reached completion.

This self-enhancing loop of reactions, a chain of positive feedbacks, is illustrated in Figure 6.10. The production of E_2 is bootstrapped by stimulating the formation of E_1 from a supply of Z_1 which, in turn, activates a supply of Z_2 to form even more E_2. The ability of E_2 to catalyze its own formation ensures that a large supply of clotting proteins is rapidly produced. However, to safeguard against an overzealous

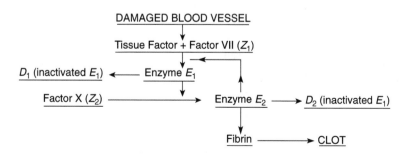

Figure 6.10. Schematic of the biochemical reactions leading to a blood clot.

production of clots, both E_1 and E_2 are quickly inhibited either by binding with other blood proteins called inhibitors or by a process of negative feedback in which E_2 indirectly acts to inactivate E_1 and, therefore, itself. The in-activated products are denoted by D_1 and D_2 in the schematic of Figure 6.10. The delicate interplay between activation and inhibition results in an initial burst of clotting followed by a rapid decline in the levels of all the enzymes involved in coagulation. This leads to what is in effect a system "shutdown" when the job is done.

The needed proteins are in plentiful supply in the blood plasma and clotting will initiate whenever they are activated. We can get a bit colloquial here and say that the clotting system is "idling" prior to an injury. There is a threshold below which the clotting cascade is not triggered into action, since small enough tissue damage should not result in large clot formation because vagrant clots can lead to a fatal thrombosis. After all, blood vessels are being ruptured every time we knock against something, however slightly, and an all-out response is generally unwarranted. By the same token, once the threshold is exceeded the system must be stimulated into action to avoid bleeding to death.

Our goal is to write down differential equations to model the reactions that represent the initial phase of clotting, and to derive from this the threshold condition for system activation. We begin first by looking at the simplest case in which a single enzyme E catalyzes its own formation without requiring the intervention of other enzymes, a situation that actually arises in certain biochemical reactions and is known as *autocatalysis*. E combines with a protein Z (in enzyme kinetics Z is called a substrate of E) to form a complex C. This is a reversible reaction that occurs at a rate proportional to the concentrations of E and Z, with a rate constant k_+, and that can then dissociate back into its constituents at a rate proportional to the concentration of C itself, with rate constant k_-. When the reaction energy is high enough E will cleave Z to form more E at a rate k, proportional to the concentration of C. E is then inactivated into form D (for "dead") at a rate proportional to its own concentration, at rate k_1. This feedback loop is shown in Figure 6.11 and gives rise to two differential equations.

$$C' = k_+ EZ - (k_- + k)C$$
$$E' = kC - k_1 E$$
(6.15)

There is an initial amount of Z, namely, Z_0, which diminishes as the reaction progresses, but the amount of E can vary up or down. At any instant of time E

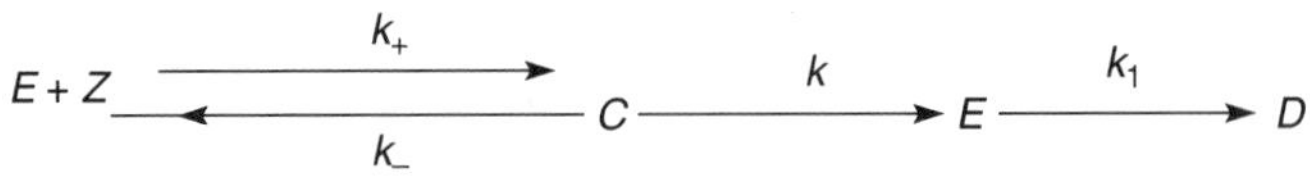

$$E + Z \xrightleftharpoons[k_-]{k_+} C \xrightarrow{k} E \xrightarrow{k_1} D$$

Figure 6.11. Schematic of the feedback loop for autocatalysis.

and Z are either freely available or they are bound up in the complex C. However, their total amounts are conserved and so we have the relations

$$Z = Z_0 - C$$
$$E = E^{\text{total}} - C \tag{6.16}$$

Substituting (6.16) into (6.15), the right-hand side of the first equation becomes a quadratic in C with roots r_1 and r_2 that are real and positive (Exercise 6.7.10); let's assume that $r_1 > r_2$. Then the first equation in (6.15) can be written in terms of its factors as

$$C' = (C - r_1)(C - r_2) \tag{6.17}$$

If one assumes that Z_0 is in large excess over the initial concentration of E during the initial phase of the reaction, which is usually the case, then the binding of E to C will take place rapidly or, to put it another way, the change in Z is "slow" compared to the "fast" formation of C. C moves quickly to the attractor r_1, where it assumes its equilibrium value (Exercise 6.7.9). After a negligibly brief period, therefore, C is essentially equal to the constant obtained by setting C' to zero:

$$C = \frac{(E^{\text{total}} - C)Z_0}{K_m}$$

where $K_m = (k_- + k)/k_+$.

Because C rapidly adjusts to any inital value of Z we may as well replace Z_0 by Z itself, whatever its value happens to be initially. Also the clumsy E^{total} is now replaced simply by E, with the understanding that E stands for the total, and time varying, amount of enzyme present in the reaction. With this convention the preceding expression for E becomes, after some simple algebra,

$$E' = \frac{kEZ}{K_m + Z} - k_1 E \tag{6.18}$$

Relation (6.18) is known as the *Michaelis-Menten equation*.

The protein Z is consumed at the same rate that E is produced and so there is an additional equation to consider:

$$Z' = \frac{-kEZ}{K_m + Z} \tag{6.19}$$

with the minus sign indicating that Z is decreasing. Some progress can be made toward solving the equation pair (6.18) and (6.19) if we notice that since Z changes very little, at least during the initial phase of the reaction, K_m is considerably greater than any change in Z that takes place. Because of this, $K_m + Z$ is essentially constant and so the equations become

$$E' = aEZ - k_1 E$$
$$Z' = -aEZ$$
(6.20)

with $a = k/(K_m + Z)$. A nullcline approach now shows that E is activated (that is, it will increase in value before eventually decaying to the attractor zero) whenever Z exceeds k_1/a. Otherwise E goes quickly to zero. Rather than pursue the consequences of this, let us return to the original problem in which there are two enzymes E_1 and E_2 (see, however, Exercise 6.7.6.)

To derive the appropriate differential equations it suffices to mimic the argument used above. To begin with, E_2 and Z_1 combine to form a complex C_1 as a reversible reaction that occurs at a rate proportional to the individual concentrations of E_2 and Z_1, with a rate constant k_+, and it dissociates back into its constituents with a rate constant k_-. When the reaction energy is high enough, E_2 will cleave Z_1 to form E_1 at a rate m_1. Then, after the cleavage is complete, E_2 is released and becomes available again. A completely parallel set of reactions leads to a complex C_2 from the combination of E_1 with Z_2. This is illustrated in the schematic of Figure 6.12 below.

We are led, as before, to an equation for the formation of C_1:

$$C_1' = k_+ Z_1 E_2 - (k_- + m_1)C_1$$

The product of this reaction, E_1, follows the equation

$$E_1' = m_1 C_1 - k_1 E_1$$

in which we assume that E_1 is inactivated to D_1 at a rate proportional to its own concentration, with proportionality constant k_1.

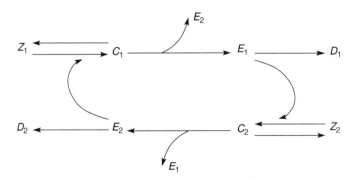

Figure 6.12. Schematic of the feedback loop for blood clot initiation involving two enzymes.

At any instant, the enzyme E_2 is available either in free form or is bound up in the complex C_1. It follows that

$$E_2 = E_2^T - C_1$$

where T stands for "total."

At this juncture we observe that during the initial stage of coagulation Z_1 and Z_2 are in large excess over E_1 and E_2 before any substantial conversion of these blood proteins has taken place. As in the simpler case of a single enzyme treated earlier, this ensures that the binding of E_2 to Z_1 (and, similarly, of E_1 to Z_2) takes place rapidly. Thus C_1 very quickly reaches a nearly constant saturation value, whereas Z_1 remains essentially constant as some value Z_{10}. This enables us to assume that C_1 is at a stable equilibrium in which its derivative is zero. Combining this with (6.18) and (6.19) results in

$$C_1 = \frac{E_2^T Z_{10}}{K_{m_1} + Z_{10}}$$

where K_{m_1} stands for the ratio $(m_1 + k_-)/k_+$. Because C_1 quickly adjusts to any initial value of Z_1 we repeat what was done earlier and replace Z_{10} by Z_1 itself, whatever its value happens to be initially. From Equation (6.19) it now follows that

$$E_1' = \frac{m_1 Z_1 E_2}{K_{m_1} + Z_1} - k_1 E_1 \tag{6.21}$$

in which E_2^T is replaced simply by E_2, with the understanding that E_2 denotes the total amount of this enzyme, with a completely analogous meaning for E_1. We therefore have a similar equation for E_2 obtained in an identical manner:

$$E_2' = \frac{m_2 Z_2 E_1}{K_{m_2} + Z_2} - k_2 E_2 \tag{6.22}$$

Since Z_1 and Z_2 are consumed at the same rate that E_1 and E_2 are produced it is also true that

$$Z_1' = \frac{-m_1 Z_1 E_2}{K_{m_1} + Z_1} \tag{6.23}$$

$$Z_2' = \frac{-m_2 Z_2 E_1}{K_{m_2} + Z_2} \tag{6.24}$$

The four Equations (6.21) through (6.24) can be linearized about the equilibrium values in which Z_1 and Z_2 are assumed constant at some values Z_{10} and Z_{20} and where E_1 and E_2 are zero. This leads (Exercise 6.7.8) to two linear equations

$$E_1' = a_1 E_2 - k_1 E_1$$
$$E_2' = a_2 E_1 - k_2 E_2 \qquad (6.25)$$

Here the constants $a_i = m_i Z_{i0}/(K_{m_i} + Z_{i0})$ are composites of the several constants appearing in the previous equations. The behavior of this system depends on the eigenvalues of the matrix

$$\mathbf{A} = \begin{pmatrix} -k_1 & a_1 \\ a_2 & -k_2 \end{pmatrix}$$

because this tells us, by virtue of Lemma 6.3, whether the equilibrium at the origin of the E_1, E_2 plane is an attractor or not. If it is, then any initial value of the enzymes will decay to zero and the system remains subthreshold, whereas an unstable equilibrium implies an explosive growth in these enzymes, signaling an initial catapulting in the amounts produced. This is the above-threshold condition that leads to clot formation. Of course, the linear equations cannot tell us what happens far from equilibrium after the initial onset of activity. For this one needs to integrate the full set of nonlinear equations and see how the system evolves over time. First, however, let us find the threshold condition by checking the eigenvalues. A quick computation based on Lemma 6.3 shows that if

$$\theta > 1$$

where $\theta = a_1 a_2/k_1 k_2$, the equilibrium is unstable (a saddle point in which one eigenvalue is positive, the other negative). This is because the trace and determinant of the matrix $\mathbf{A}$ are both negative. Otherwise, with $\theta < 1$ the equilibium is an attractor with two real and negative eigenvalues since the determinant is now positive. Note that θ is a quotient of activation to inhibition rates and it is this ratio that determines whether the clotting system will ignite or not. A large enough injury will release a sufficiently large amount of tissue factor, and coagulation then gets off the ground by guaranteeing that θ will exceed unity. If the inhibition rates k_i are large enough, for example, then no clotting will take place. Patients with severe clotting problems are sometimes treated with an infusion of the anticoagulant drug heparin that actually increases these rates. An oral "blood thinning" drug taken by heart-attack victims, warfarin, achieves similar results by decreasing the constants m_i.

An integration of Equations (6.21) to (6.24) is shown in Figure 6.13 in which there is initially a small amount of E_1 present to get the process started. The two plots show E_1 under sub- and above-threshold conditions. In the first instance, E_1 decays to zero, and in the second, we see a substantial rise in its concentration as a result of activation, but with an eventual decay to zero again because of inhibition.

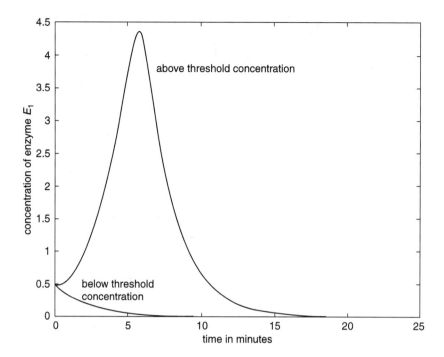

Figure 6.13. Threshold response for the activation of enzyme E_1, from the blood clot model Equations (6.21) through (6.24).

In Chapter 7 we will come to think of this kind of threshold behavior as characteristic of a wide class of living systems that exhibit excitability in response to a large enough stimulus.

6.7. Exercises

6.7.1. Specific solutions to scalar equations of the form $x' = f(x)$ can sometimes be obtained by the following procedure. Bring $f(x)$ over to the left side of the equation and then integrate both sides with respect to the variable t:

$$\int f(x(t))x'(t)\,dt = t + \text{constant}$$

By the change of variable formula for integrals, the left side can be written as $\int f(x)\,dx$ and if this integral can be evaluated simply then, quite often, x will be expressed explicitly in terms of t. An example is

$$x' = rx(1 - x)$$

which can be written as

$$\frac{x'}{x(1-x)} = r$$

Integrating as already indicated gives $\log x - \log(1-x) = rt + \text{constant}$. (Note that $1/(x(1-x)) = 1/x + 1/(1-x)$. After a little bit of algebraic manipulation one gets

$$x(t) = \left(1 + \frac{e^{-rt}}{r}\right)^{-1}$$

which shows that $x(t) \to 1$ as $t \to \infty$.

Solve Equation (6.5) using the same method and then show that $x(t)$ tends to $K(1 - E/r)$ as t goes to infinity.

6.7.2. Do a nullcline analysis on model Equations (6.6) for the case $r/a > L$, $s/b > K$. What does this suggest about the non-trivial equilibrium ?

6.7.3. Establish Lemma 6.3, namely, that if Det $\mathbf{A} > 0$ and Tr $\mathbf{A} < 0$, then the real parts of the eigenvalues of the Jacobian are negative, whereas if Det $\mathbf{A} < 0$, regardless of the sign of the trace, then the eigenvalues are real and of opposite sign. Hint: the eigenvalues of the Jacobian are roots of a quadratic equation (also known as the characteristic equation). Solve this equation explicitly.

6.7.4. Two enemy forces engage in guerilla warfare in which they annihilate each other at rates proportional to the product of the number of troops on each side of the conflict. This is a reasonable assumption in view of the fact that encounters between them become more frequent as their respective sizes increase. The rates need not be equal because one force may be more effective than the other. Model this by a pair of equations and show, depending on the initial troop sizes, that one side or the other will be eliminated in finite time. This is the simplest form of what is known as a combat model. How would you modify the equations in the event that one of the two sides suffered casualties from self-inflicted wounds at a rate proportional to the number of troops remaining?

6.7.5. Do a nullcline approach to Equations (6.14) to deduce oscillatory behavior of the solutions about the equilibrium. Use Lemmas 6.2 and 6.3 to arrive at the same conclusion.

6.7.6. Equations (6.14) can be simplified further by ignoring the birth rate r. This supposes that the population size is fixed with no new recruits or defections during the period of an epidemic, which is a reasonable assumption for diseases of short duration. The equations now read as

$$S' = -bSI$$
$$I' = bSI - cI$$

(6.26)

The fraction R of recovered individuals is initially zero and, because $S + I + R = 1$, this means that $S + I = 0$ at $t = 0$. From the chain rule of differentiation one obtains, for all S, I in the positive quadrant,

$$\frac{dI}{dc} = \frac{I'}{S'} = -1 + \frac{c}{bS}$$

Integrating both sides of this relation with respect to S yields

$$I(t) = -S(t) + \frac{c}{b} \log S(t) + \text{constant} \tag{6.27}$$

Assume that $c/b < 1$. Plot the solution curves in the S, I plane and interpret the results in light of the discussion in Section 6.4. Show, in particular, that if S is initially greater than c/b, there is an epidemic, after some infectives are introduced into the population, whereas if S begins at some value less than c/b the disease doesn't take root and the infectives diminish over time. This is the threshold condition. In any case, show that the epidemic dies out not for any lack of susceptibles in the population but rather because there are no more infectives.

Note that the equation pair (6.26) is formally equivalent to the model Equations (6.20) for autocatalysis in blood clotting. Indeed both models observe a threshold condition for growth and decay, suggesting similarities in the underlying dynamics.

6.7.7. Differentiate the first-order linear equations (6.25) for E_1 and E_2 to obtain a single second-order differential equation. Now let $u = E_1'/E_1$ to obtain a first-order nonlinear equation known as a *Riccati equation*. Apply Lemma 6.1 to this scalar equation to deduce stability or instability of the equilibria and compare to the results obtained in the text.

6.7.8. Carry out the linearization of (6.21)–(6.24) to obtain (6.25).

6.7.9. Solve Equation (6.17) explicitly using the method described in exercise 6.7.1 and show that $C(t)$ tends to r_2 as t goes to infinity. First, however, establish the roots r_1 and r_2 of the quadratic right side of this equation are real and positive.

6.8. Further Readings

The models discussed in this chapter have their origins in literature going back many decades and are, by now, classical. An excellent introduction to models of interacting populations is to be found in the collection of essays [15] and the comprehensive book [26], while epidemic models are the subject of the books [2] and [4].

The necessary background in differential equations can be acquired from the book by Jeffrey [31], but since we also rely on numerically computed solutions it would be useful to have available a software package for PC use that displays the solutions graphically, as we often do in the text. MATLAB, Version 5, is used throughout this book.

After the vaccination of school children began in earnest in many communities around 1963, the incidence of measles declined significantly only to reappear among preschoolers during the late 1980s (see, for example, "Resurgence of Measles Prompts New Recommendations on Vaccination," *New York Times*, Jan. 11, 1990). The idea that some recurrent epidemics are a manifestation of chaotic dynamics is based on the paper [44] and is reviewed in a short note in *Science* [45]. An elementary introduction to chaotic attractors is to be found in the *Scientific American* article [25].

The threshold condition for blood clot initiation that was discussed in Section 6.6 is from the paper by Beltrami and Jesty [8].

We cannot resist quoting a few lines from Dr. Seuss's book *On Beyond Zebra!* (Geisel, 1955) in which he talks about creatures called Nutches who are in competition with each other. The quote is especially apt in view of the discussion in Section 6.2:

These Nutches have troubles, the biggest of which is
The fact there are many more Nutches than Nitches.
Each Nutch in a Nitch knows that some other Nutch
Would like to move into his Nitch very much.
So each Nutch in a Nitch has to watch that small Nitch
Or Nutches who haven't got Nitches will snitch.

(From ON BEYOND ZEBRA! by Dr. Suess, TM & copyright © by Dr. Suess Enterprises, L.P. 1955, renewed 1983. Used by permission of Random House Children's Books, a division of Random House, Inc.)

Sardines and Algae Blooms

7.1. Background

In his novel *Cannery Row* John Steinbeck vividly describes the bustle of a sardine factory in Monterey, California, before the Second World War, when heavily laden boats brought in the abundant catch of the sea. But the industry suffered a sharp decline as the sardine population collapsed in the late 1940s and the once thriving canneries began to close down. Later, in the early 1970s a similarly precipitous drop in the catch of anchovies occurred in Peru, the then largest fishery in the world. What happened? Although some episodic shift in climate may have had an influence, especially the influx of warm waters known as "El Niño," which occurs sporadically off the Peruvian coast, the dominant factor in each instance was, quite simply, overfishing.

When commercial fishing begins along some coastal region this sets in motion an investment in fishing gear that continues to increase to exploit what initially appears to be an ample natural resource and within a few years a sizable fleet plies the waters. As fish stocks dwindle and prices rise, the competition among fishermen intensifies and even more aggressive fishing tactics may be the response. Certain species of fish, such as sardines, anchovies, and herring, swim in schools that not only favor their detection and bulk harvest but, in biological terms, their ability to breed may be diminished by low population densities. This implies that an overly exploited species can have a diminished capacity to recruit new adults. In Section 7.2 a model is presented that combines both economic and biological considerations and that exhibits how the sudden collapse of just such a species becomes possible as a critical level of harvesting is reached.

Episodes of algae growth have been recorded in estuaries and coastal areas for several hundred years and most have been considered to be benign events or, at worst, mild nuisances that recur with humdrum regularity in late spring and summer. However, in recent years there have been more sporadic, less predictable, and certainly more severe bouts of elevated cell growth of certain toxic algae species, known generically as red tides or even brown tides. There is a mounting worldwide concern for these unusual bloom episodes because of their adverse impact on fisheries and aquaculture and from the eutrophication that results from their collapse. These harmful marine flare-ups are marked by a sudden proliferation of cell counts followed, quite often, by rapid disintegration.

The onset of the outbreaks can be attributed, in part, to a confluence of climatic and meteorological conditions that alter the physical and chemical composition of the waters in such a way as to predispose one algal species to achieve dominance in preference to other species that normally inhabit the same territory. The intruder may arrive as a castaway in ship ballast or from a storm event far out at sea and it seizes the opportunity for unbridled growth when the local conditions are favorable. However, even if bloom initiation is largely due to chance events, its subsequent duration and severity are moderated by other organisms, the zooplankton, that graze on the algae. We will discuss a simple model of this prey-grazer interaction that offers a plausible metaphor for the observed dynamics of a bloom in progress.

A similar situation prevails in the spruce and fir forests of eastern Canada and the northeastern United States. A patch of forest is suddenly defoliated by an explosive increase in the numbers of a voracious caterpillar called the spruce budworm, which denude the forest patch within several consecutive years. This is followed by the sudden decrease in the number of budworms. A forest canopy has now been opened up that permits new seedlings to grow and the forest renews itself. Air currents allow budworm moths to migrate from other forest patches, however, and eventually the setting is ripe for a new outbreak to occur.

What algae blooms and insect outbreaks have in common is that these events are triggered whenever a certain threshold in favorable conditions is attained, such as a high enough salinity level for the algae or a minimum density of fir and spruce needles for the insect. This phenonemon is reminiscent of the thresholds needed for the spread of infectious diseases or the activation of an enzyme cascade that leads to blood clotting, ideas that we first encountered in the previous chapter. In effect, these are all *excitable systems* that remain quiescent until suddenly ignited by propitious events.

The stimulus to changes in the natural world are not always sporadic and unpredictable, as in the case of unusual blooms. Cyclic or nearly cyclic phenomena abound in nature as biological systems respond to periodic fluctuations in the environment around them. One thinks of daily rhythms that are "tuned in" to night following day following night, or of bodily rhythms as well as insect and plant behavior that are influenced by lunar and seasonal cycles. Periodic orbits can also come about from the interplay between activation and inhibition. In the fish-harvesting model, for example, the self-enhancing growth of fish is balanced by exploitation and this gives rise to cyclic behavior, as we will see in Section 7.5, where we take a second and quite different look at the model.

7.2. A Catastrophe Model of Fishing

It appears that sardines and similar species of fish have a difficult time breeding at low population densities and their survival is enhanced with increasing density, up to a level at which overcrowding begins to have an inhibitory effect on reproduction. This is achieved by swimming in "schools," large and closely packed swarms, because this confers more protection from predators. Unlike some other species that follow a logistic per capita growth rate $r(1 - x/K)$, which decreases linearly with an increase in density x, the per capita growth rate of sardines is probably better modeled by a term that begins low and increases with x for a while before decreasing. This would reflect an initially low growth rate that increases with population density x and then declines when x is large. A simple example of this is the function $rx(1 - x/k)$.

Consider now a fishing zone with unrestricted access. Any number of fishermen, with their trawlers and nets and, in recent years, electronic fish detectors, can harvest the fish. Let E be the total fishing effort in terms of the vessels and fishing gear, as well as manpower deployed per unit time, and let v denote the fraction of each ton of fish that is caught per unit effort. Then vEx is the total catch per unit time, where x is the fish density in the fixed zone, measured in tons. An increase in E signifies that fishing effort has intensified. If we adopt the per capita reproduction rate of schooling fish then the rate of change of x is given by

$$x' = rx^2\left(1 - \frac{x}{K}\right) - vEx \tag{7.1}$$

where the second term on the right reflects the loss due to fishing. We note, parenthetically, that if the per capita reproduction rate decreases linearly with x, (7.1) becomes the logistic Equation (6.4) of the previous chapter.

Now we introduce some drastically simplified market economics. Suppose that the cost, in dollars, per unit effort, is c and that p is the price obtained for the fish at the pier, in dollars per unit catch. The cost includes operating expenses as well as an amortized capital investment, while p reflects the market value of the fish. The net revenue from the harvest is proportional to $pvEx - cE$. As long as this quantity is positive, the fishing effort increases. This follows from the idea that when a natural resource has unrestricted access its exploitation continues unabated until the resource no longer provides any profit. In the case of a fishery the harvesting would increase until the net revenue is zero or until the existing fish stock is exhausted. A fisherman who does not participate in the haul relinquishes his portion of the take to his competitors. A sole owner of the fishery might want to harvest less relentlessly, because a prematurely depleted fishery does him no good, and he has the incentive to balance a short-term gain from aggressive fishing against the potential long-term benefits of gradual exploitation. But in an open access fishery, the short-term myopic view prevails that as long as a profit is to be made there is no reason to incur a loss by not joining the scramble for what is left.

Of course, this is a gross simplification because no fishery is totally unregulated and, as fish stocks begin to be depleted, a conservationist attitude becomes more prevalent among the regulating agencies. Moreover, as competition intensifies for a dwindling resource, the effort must necessarily increase, which would discourage those fishermen who have alternative employment opportunities. Nevertheless we adopt this somewhat fictional view and write an equation for the rate of change of fishing effort as

$$E' = \alpha E(pvx - c) \tag{7.2}$$

This says that E increases or decreases at a rate proportional to the net revenue with a constant of proportionality α. If the net profit is positive, it increases; otherwise it decreases. This constant of proportionality is generally small to reflect the fact that there is a certain inertia in the way the fishing industry responds to a perceived change in either market conditions or in the availability of fish in the ocean. It takes some time to hire or lay off personnel, invest in new gear, or to put additional boats out to sea. These changes take place at a smaller pace than the breeding of fish. For this reason E is considered to be a *slow variable* relative to x, which is deemed to be *fast*.

At this juncture we need to modify Equation (7.1) by adjoining a small constant e:

$$x' = e + rx^2 \left(1 - \frac{x}{K}\right) - vEx \tag{7.3}$$

The interpretation of the constant e is that even when the observable population density x is zero some additional fish are spawned because a few members of the species found a refuge, or because additional fish migrate into the fishing zone at some constant rate. We are considering a threshold effect here that just barely allows for the possibility of recovery. It also embodies the idea that when fish stocks are sufficiently reduced, further exploitation ceases either because it isn't worth the effort or because a regulatory agency has imposed a ban.

Our fishery model consists of the coupled Equations (7.2) and (7.3).

We begin the discussion by noting that the scalar Equation (7.3) has either one or three equilibria, depending on the value of E, which are obtained by setting x' to zero. This occurs at the intersection of the curve $g(x)/vx$ with the line defined by $E = $ constant, in the x, E plane, where $g(x) = e + rx^2(1 - x/K)$. This is illustrated in the six panels of Figure 7.1 for different values of E.

By examining the sign of x' we can determine the stability properties of the equilibria, as explained in Section 6.2. A single equilibrium is always an attractor but if one has three equilibria the middle one is always unstable. These results follow from the fact that x increases or decreases according to whether the value of E is less or greater than $g(x)/vx$.

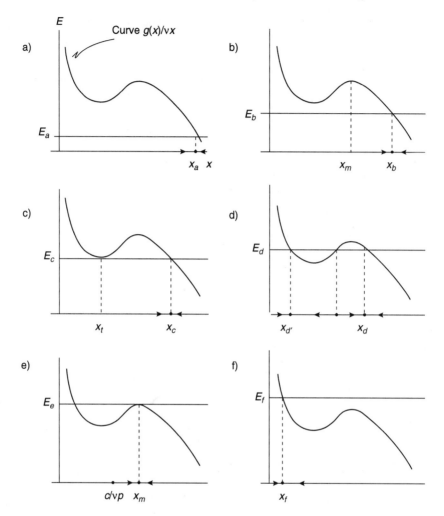

Figure 7.1. The curves $g(x)/vx$ and $E = $ constant, for six different values of E, showing the equilibrium values of the fish stock x as obtained from model Equation (7.3). E is treated as a slowly changing parameter relative to the fast variable x and the six panels are snapshots of how fish stocks change in response to changing values of harvesting rates.

Because E varies slowly with respect to x (recall that α is a small constant), it follows that x responds quickly to a change in E and so it resides at or near its closest attractor. This means that a study of the pair of equations for x and E can be simplified considerably by letting E be a gradually changing parameter in the single Equation (7.3) for x. The fish level then corresponds to the value of x that occurs at the intersection of E with $g(x)/vx$.

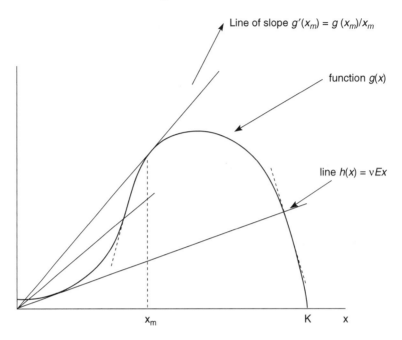

Figure 7.2. Intersection of the curve $g(x)$ and straight line $h(x) = vEx$ for different values of E.

Consider now the following scenario. Initially the fishing effort is slight, say E_a, and so a single equilibrium occurs at a high value of x, indicated by x_a in Figure 7.1a. This is an attractor, as we already noted, and so the fish stock is x_a when fishing effort is E_a. A modest increase in E to E_b reduces the equilibrium to x_b but, even though fish density x is now less, the total yield vEx is greater (Figure 7.1b). To see why this is so, observe that the function $g(x)/vx$ achieves its local maximum at the hump shown in Figure 7.1b at the point x_m. The derivative of this function is therefore zero there and its second derivative is negative. It can be shown, in fact (the computational details are shown later, in Section 7.5), that $g(x)/x = g'(x)$ at x_m. Now the yield, namely the function $h(x) = vEx$, is a straight line that intersects the function $g(x)$ and, as shown in Figure 7.2, the value of $h(x)$ at the point of intersection at the point x_m is where $g(x)/x$ and, therefore, the yield attains its maximum. The corresponding value of E, namely E_m, is called the *maximum sustainable yield*. Until E reaches the value E_m, the yield continues to increase.

As E increases beyond E_b there is a value E_t at which the line $E = $ constant is first tangent to the curve $g(x)/vx$ (Figure 7.1c). The point of tangency occurs at $x = x_t$ and is called a *bifurcation point* since two new equilibria are created for all E that are just beyond this point of tangency. However, the equilibrium at

x_c is still an attractor. There are three equilibria when $E = E_d$, as can be seen in Figure 7.1d, but the middle one is unstable, and because x moves quickly to the nearest attractor, the fish density resides at x_d and does not move to the other attractor at $x_{d'}$.

Beyond this level of harvesting effort the fish stocks continue to decline and prices begin to rise through scarcity. Ironically this can draw new entrants into the fishery and the competition to catch intensifies as long as $\frac{c}{vp}$ is less than x_m, where $\frac{g(x)}{vx}$ attains its local maximum, since this implies that the derivative E' in Equation (7.2) remains positive. As E increases beyond E_d we enter a phase of biological overfishing since immature fish begin to be captured before they can breed and eventually, when the point of tangency at x_m is attained, there is another bifurcation at which two of the equilibria coalesce and disappear (Figure 7.1e), beyond which there is a single equilibrium at a much lower fish density x_f where $E = E_f$. This is an attractor and so the fish stocks suddenly fall to this new equilibrium, an event that signifies the collapse of the fishery (Figure 7.1f).

Assume now that the ratio $\frac{c}{vp}$ is greater than x_f. We see from Equation (7.2) that the derivative E' is now negative and so fishing effort reverses itself. In effect it no longer pays to go out to sea and many trawlers begin to rust in the harbor.

Eventually recovery of the fish stocks takes place because of reduced harvesting. As E decreases, the situation pictured in Figure 7.1d eventually reappears but now the fish density moves to its nearest attractor, namely, $x_{d'}$. By reducing E further, there is ultimately only one attractor, at the relatively high value just beyond x_c (Figure 7.1c). The fish density therefore makes a sudden resurgence at a level that exceeds $\frac{c}{vp}$, where the derivative of E is once again positive. We assume here that $\frac{c}{vp}$ is between x_t and x_m; otherwise, for $\frac{c}{vp}$ less than x_t, fish recovery is less dramatic than indicated in the present scenario.

Some ships feel encouraged to go out to sea and the process starts over again, leading to a cyclic boom and bust population dynamic. A *hysteresis* effect takes place in which the path to recovery follows a different route than the path to collapse (Figure 7.3). When Equations (7.2) and (7.3) are considered simultaneously as a pair, the nullcline method that was explained in Section 6.1 suggests that a closed cycle does indeed exist in the x, E plane. A more complete discussion of this cyclic behavior is reserved for Section 7.5.

We assumed throughout that K is constant but if this quantity is allowed to vary in response to environmental fluctuations that affect the maximum sustainable population density of the fish, then the equilibrium values of fish stock x depend on K as well as E. The function $\frac{g(x)}{vx}$ must now be replaced by $\frac{g(x,K)}{vx}$, where the dependence on K is made explicit. Figure 7.4 illustrates this two-dimensional function in which $g(x)/vx$ are simply slices through this surface for different values of K.

Models in which rapid jumps take place from one plateau to another as a certain parameter crosses a bifurcation point form part of *catastrophe theory*, which explains the title of this section. Note that if the constant e is zero, the population crashes to zero and becomes extinct. On the other hand, if the per capita growth

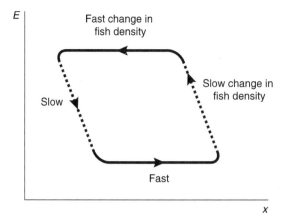

Figure 7.3. Stable equilibrium values of the variable x in response to changes in E. As fishing effort E increases and then decreases a hysteresis effect is observed in which fish density x follows a different path from collapse to recovery.

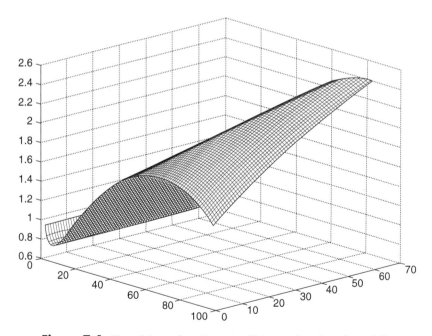

Figure 7.4. Plot of the surface $E = g(x, K)/x$ as a function of x and K.

is logistic, then a quite different dynamic takes place in which x is asymptotically stable at the value of $c/\nu p$. This can be demonstrated in a similar way, and there is no need to repeat the nearly identical arguments here.

We have attributed a crash in fish density to overfishing. There could also be unfavorable climatic changes that restrict the rate at which fish larvae develop into young adults. This corresponds to a decrease in the rate constant r. These are called recruitment failures and an already seriously depleted fishery is especially prone to collapse when an environmental fluctuation such as "El Niño" occurs. Moreover, a diminished species can be replaced by other, commercially uninteresting, species and these may inhibit the recovery of the original fishery. We saw how this can happen in the preceding section when one of two strongly competing species drives the other to extinction as we move from one basin of attraction to another.

The present model assumes that there is no seasonal preference for hatching and that reproduction is unaffected by age and sex differences. In spite of these and similar simplifications the model appears to mimic the gross behavior of certain types of fisheries, such as anchovies and sardines, in which excessive exploitation can lead to a sudden collapse followed by gradual recovery.

A dramatic example of such a collapse is provided by the Peruvian anchovy fishery. Using the number of vessels deployed per unit time as a measure of effort, the actual catch (in 10^6 tons) was given by

Year	Effort	Yield
1959	1.4	1.91
1964	5.8	8.86
1968	8.9	10.26
1973	46.5	1.78

As E increased from 1963 so did the catch but by 1973 the yield had fallen to pre-1959 levels.

If the cost to price ratio c/p increases to a level at which $c/\nu p$ exceeds x_m the fishery avoids a crash (Exercise 7.6.1). This can be ensured by governmental intervention in the form of imposing a tax T per unit catch. This effectively reduces the net price for the fish to $p - T$ and, if the tax is severe enough, $c/\nu(p - T)$ will indeed exceed x_m. There are other regulations that can be and are imposed on fisheries, especially after a disastrous failure, such as catch quotas, licenses and fees that restrict the entry of additional fishermen, and no fishing zones. These are not considered here.

7.3. Unusual Blooms

Since 1985 a small (2–3 μm in diameter) algae called *Aureococcus anophageffer-ens* or, simply, *A.a.*, has sporadically erupted in the Peconic estuary and nearby embayments of eastern Long Island, New York. The commercial harvesting of scallops and mussels is seriously disrupted during one of these blooms because the shellfish cannot ingest *A.a.* and they therefore starve. Cell counts of *A.a.* catapult within days to over 10^6 cells/liter, and even as high as 10^9 in at least one year, leaving the waters a impenetrable muddy haze that is called a "brown tide."

Certain microzooplankton, small crustaceans, feed on *A.a.* and other tiny algae that are generically classified as phytoplankton. This trophic link between the zooplankton and their prey can be modeled, in the simplest case, by a pair of differential equations in the variables x and y, where x is the concentration of *A.a.*, in units of 10^5 cells per liter, and y is the concentration of the grazing zooplankton in units of 10^3 per liter. The equations are written first and then the individual terms are described in more detail:

$$x' = rx\left(1 - \frac{x}{k}\right) - \frac{\gamma x^2 y}{\delta + x^2}$$

$$y' = \frac{\varepsilon \beta x^2 y}{\delta + x^2} + dy - cy^2$$

(7.4)

Growth rate of the prey is taken to be logistic, as in the population models of Chapter 6, with a theoretically maximum sustainable population density of k that ranges from 20 to 30 (in units of 10^5 cells per liter, as already noted) and the per capita growth rate r is calibrated to correspond to one to two cell divisions per day at the height of a bloom. The value of k will change according to local environmental factors that serve to either trigger or inhibit a bloom. This will also be true of r because the cell-division rate is temperature dependent and therefore varies with the climate.

The grazing rate is estimated at about 0.5 cells per liter per day, a figure used in setting the predation rate γ in the model. The quantity β is a fraction that measures the conversion of ingested cells into zooplankton biomass and is roughly 0.4.

The quantity ε is a small constant to ensure a correspondingly small variation in y compared to x. This says that changes in y are "slow" compared to "fast" changes in x.

The form of the predation term requires some explanation. At the very beginning of a bloom, grazing is negligible because of the availability of alternative and preferable food sources. But, as the bloom progresses, the zooplankton shift their attention to *A.a.* because it is becoming an abundant and even dominant species. Also, at high population densities, there is an increasing rate of algal mortality caused by viruses that lyse the cells and hasten bloom disintegration. This means the growth is proportional to x^2 instead of just x because x^2 increases slowly at first and much more rapidly when x is at elevated levels. The denominator serves to measure satiation of the grazer as x becomes abundant: when x is large the predation term is essentially proportional to y alone, whereas for small to moderate x, it depends on the interaction between both predator and prey. As the parameter δ decreases, satiation takes place more rapidly.

The quantity d is a per capita growth rate of grazers due to the availability of alternative food sources, and cy^2 is a death rate that increases with y^2 to signal the combined effects of increased predation by larger "macrozooplankton," such as copepods, on the smaller zooplankton as the latter become more abundant, as well as more aggressive competition and interference among the various species of feeding microzooplankton. In effect, this term is a surrogate for a three-tiered trophic chain that includes two different levels of predators. The parameters δ, d, and c were calibrated to give computational results that are consonant with known observations of actual bloom events.

Within the chosen range of parameters there is a sole equilibrium, call it x_{equil}, for Equations (7.4), which is the intersection of the x and y nullclines. Figure 7.5 shows the nullclines. From this figure and the first of Equations (7.4), we see that if y is above the x-nullcline, then x' is negative; when y is below the nullcline, x' is positive. We therefore conclude that the equilibrium is an attractor.

We come now to the crucial point of our discussion. Even though x_{equil} has been identified as an attractor, so that all trajectories in its vicinity will eventually wind

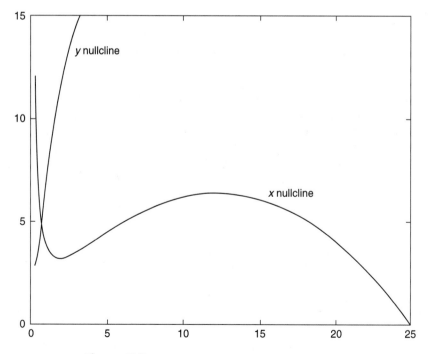

Figure 7.5. Nullclines for the model Equations (7.4).

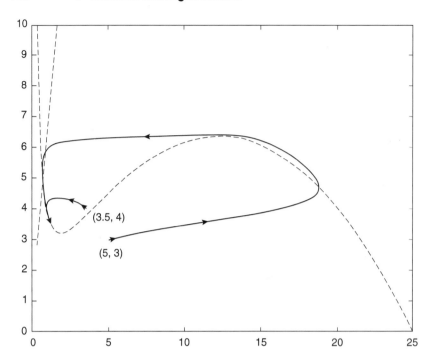

Figure 7.6. Trajectories for the algal model Equations (7.4) for two nearby starting values of the variable x, with nullclines shown as dotted lines.

their way down to it, the path taken may vary enormously depending on where the trajectory begins. The pivotal idea here is that two solutions that begin close together can have widely diverging orbits. One of them will make a large excursion before damping down to the equilibrium and the other will barely move at all. This phenomenon is called *excitability*: as the model parameters r and k vary, there is a *threshold* for these values below which a solution to the equations will remain relatively quiescent, but will suddenly blossom into a large orbit when the threshold is exceeded. In biological terms we think of the system as inhibited in one case and activated in the other, not unlike the blood clotting model of Section 6.6.

In Figure 7.6 two trajectories are plotted that begin nearby. We imagine that a small shock has perturbed the parameter values in such a way that the solution begins at the slightly displaced location shown in the figure. This position is below the x-nullcline and so the ensuing orbit will move to the right until it crosses over to above the nullcline and returns to the equilibrium. Thus, the trajectory damps down to the attractor in one instance and is excited into a large orbit in the other, which shows that exceeding a threshold can trigger a rapid proliferation in the values of x and y. This is an algae bloom.

In terms of our model, excitability means that there is a rapid jump from a low endemic level of $A.a.$, when this nuisance organism is merely a background species, to a surge in cell counts. After the bloom is initiated there is a refractory period during which high cell counts persist for a spell even if the conditions that ignited the bloom have subsided. In terms of Figure 7.6, this means that once the trajectory has moved far enough to the right, a sudden change in a parameter value back to its subthreshold level will have little effect on the subsequent path followed by the trajectory; it will continue on a wide arc until it eventually declines rapidly toward the equilibrium. To put it another way, once the "die is cast," there is enough inertia in the system to allow the bloom to sustain itself even if the triggering agents have, in the interim, fallen to subthreshold levels. In the Peconic estuary, for example, an increase in the temperature and salinity of the water, which have the effect of raising the values of r and k, are implicated as factors that elevate cell counts of *A.a.* If a threshold is exceeded and a bloom ignited, then, as the high cell counts progress through weeks and even months into the late summer, it will often happen that both temperature and salinity have already waned. The protracted nature of the algal outbreak appears to be a manifestation of the refractory behavior exhibited by the model.

7.4. Cycles

In Chapter 6 we encountered the idea of a cyclic attractor (periodic trajectory). We now discuss a theorem that tells us when such attractors may be expected. This is called the *Hopf bifurcation theorem*, which is stated, without proof, for planar systems of differential equations. There is a corresponding result for systems in R^n but this will not be needed.

As in Section 6.2, we treat a system of differential equations whose solutions are vector-valued functions of time t, except that now we allow the dependency on some scalar parameter s:

$$\mathbf{x}' = \mathbf{f}(\mathbf{x}, s) \tag{7.5}$$

Our previous discussion has conditioned us to expect bifurcations to occur as the slowly varying s attains certain critical values.

This will again be the case but, in contrast to the bifurcations encountered earlier, we now seek periodic solutions.

Let $\mathbf{x}_s$ denote an equilibrium solution to (7.5), in which the dependency on s is made explicit. The Jacobian matrix of the linearized equations will also depend on s and we write the linearized system in vector form as

$$\mathbf{u}' = \mathbf{A}(s)\mathbf{u} \tag{7.6}$$

It will be assumed that the eigenvalues λ of $\mathbf{A}(s)$ are complex:

$$\lambda_1(s) = a(s) + ib(s)$$
$$\lambda_2(s) = a(s) - ib(s) \tag{7.7}$$

where the real and imaginary parts exhibit dependency on *s* as well. The following result is then valid under the usual conditions on differentiability.

Lemma 7.1   *Assume that* $\mathbf{x}_s$ *is an attractor for* $s \leq 0$ *and a repeller for* $s > 0$. *The parameter value* $s = 0$ *is called a Hopf bifurcation point if* $a(0) = 0, b(0)$ *is nonzero, and the derivative of* $a(s)$ *is positive when evaluated at* $s = 0$. *Under these conditions, there is a range of positive s values for which there will exist a periodic solution, with the amplitude of this trajectory increasing with s.*

In view of Lemmas 6.1 and 6.2, it suffices to assume that $a(s)$ *is negative (positive) when s is negative (positive) to ensure that* $\mathbf{x}_s$ *is an attractor for* $s < 0$ *and a repeller for* $s > 0$. *However, the Lemmas tell us nothing about what happens when* $s = 0$ *because the real part of the eigenvalues is zero there. This is a delicate issue that needs to be resolved in each instance by referring to the nonlinear equations directly or, quite simply, by assuming that* $\mathbf{x}_s$ *remains an attractor at* $s = 0$ *based on extraneous evidence supplied by our observations of the problem being modeled.*

A prototypical example is provided by the equations

$$x' = sx + y - x(x^2 + y^2)$$
$$y' = -x + sy - y(x^2 + y^2)$$

in which we let $x_1 = x$ *and* $x_2 = y$ *for simplicity of notation. The sole equilibrium occurs at the origin in the* x, y *plane and the Jacobian matrix of the linearized system at the equilibrium is*

$$\begin{pmatrix} s & 1 \\ -1 & s \end{pmatrix}$$

The determinant of this matrix is always positive but the trace $2s$ *can be of any sign. The eigenvalues are* $s + i$ *and* $s - i$, *and so* $a(s) = s$ *is zero at* $s = 0$, *whereas* $b(s)$ *is the nonzero imaginary. The real part* $a(s)$ *is simply trace/2 and its derivative with respect to s is 1. To see what happens when* $s = 0$ *make a change of variables in the equations by letting* $x = r(\sin \theta z)$ *and* $y = r(\cos \theta z)$, *where r represents radius in polar coordinates and* θ *is the angle. After a bit of algebra the equations are transformed into the pair*

$$r' = r(s - r^2)$$
$$\theta' = -1$$

(7.8)

The second equation says that the angle is turning clockwise at a constant rate, whereas an application of Lemma 6.1 tells us that when s is zero the scalar equation in r has an attractor at $r = 0$. *In fact one can integrate* $r' = -r^3$ *using the method of Exercise 6.7.1 and find that* $r(t)$ *tends to zero as t goes to infinity (assuming* $r(0) > 0$). *Thus the origin is in fact asymptotically stable for* $s \leq 0$ *and so all the conditions of Lemma 7.1 are now satisfied. We conclude that there must be a*

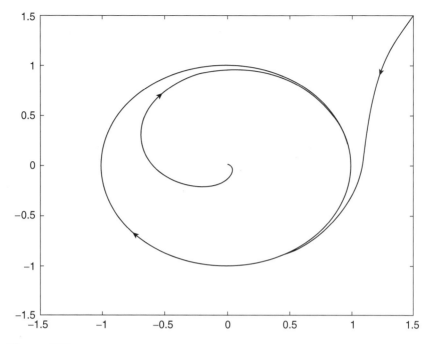

Figure 7.7. Integration of the model Equations (7.8) showing trajectories that spiral out to a periodic attractor.

periodic solution when s > 0. In fact an examination of the sign of the derivative r′ in (7.8) immediately shows that when s > 0 the origin is a repeller while the circle of radius r = $\sqrt{s}$ is an attractor (equivalently, one can invoke Lemma 6.1). Hence the trajectories spiral outward from the origin toward the circle of radius $\sqrt{s}$ and inward from positions beyond the circle. Moreover the radius of this periodic attractor increases with s. This is illustrated in the Figure 7.7.

7.5. Another View of Fish Harvesting

An interesting application of Lemma 7.1 is to the fishery model of Section 7.2 because it leads to an alternate formulation of the "boom-and-bust" scenario that comes in the wake of overfishing. Referring to the discussion in that section, recall that the equations are written as

$$x' = g(x) - \nu Ex$$
$$E' = \alpha E(\nu px - c)$$

where $g(x) = e + rx^2(1 - x/K)$. Assuming that $E > 0$, the sole equilibrium occurs at $x_\star = c/\nu p$ and $E_\star = g(x_\star)/x_\star$.

The quantity $g(x)/x$ represents the per capita growth rate and it has a maximum at some point x_m (see Figure 7.1b). The derivative of $g(x)/x$ at x_m is therefore zero and so

$$\left(\frac{g(x)}{x}\right)' = \frac{g'(x)}{x} - \frac{g(x)}{x^2} = 0$$

at x_m. This implies that

$$g'(x_m) = \frac{g(x_{\max})}{x_{\max}} \tag{7.9}$$

Moreover, because $g(x)/x$ is maximized at x_m, its second derivative there is negative:

$$\left(\frac{g(x)}{x}\right)'' = \frac{g''(x)}{x} - \frac{g'(x)}{x^2} - \frac{g'(x)}{x^2} + \frac{2g(x)}{x^3} < 0$$

and, in view of (7.9), the last three terms cancel and we obtain

$$g''(x_m) < 0 \tag{7.10}$$

The Jacobian matrix of the linearized equations corresponding to the nonlinear system for the variables x and E is readily computed at the equilibrium of this system and is

$$\mathbf{A} = \begin{pmatrix} g'(x_\star) - g(x_\star)/x_\star & -c/p \\ \alpha p g(x_\star)/x_\star & 0 \end{pmatrix}$$

and we see that the determinant of $\mathbf{A}$ is always positive. To decide the sign of the trace $g'(x_\star) - g(x_\star)/x_\star$ refer to Figure 7.2, where we see that where the line $h(x) = vEx$ intersects the curve $g(x)$ the slope of $h(x)$ is $g(x)/x$. At $x = x_m$, however, the slope of $g(x)$ is identical to $g(x)/x$, as we see from relation (7.9). If the intersection occurs at a point where x is less than x_m, then the slope of the line is less than the slope of curve g and just the opposite is true when x exceeds x_m (Figure 7.2).

Because of these considerations, the trace of the Jacobian of $\mathbf{A}$, namely, $g'(x_\star) - g(x_\star)/x_\star$ is positive when $x_\star < x_m$, zero if $x_\star = x_m$, and negative otherwise. Let $s = x_m - x_\star$ be a bifurcation parameter. Then

$$\text{trace } \mathbf{A} < 0 \quad \text{if } s < 0$$

$$\text{trace } \mathbf{A} = 0 \quad \text{if } s = 0$$

$$\text{trace } \mathbf{A} > 0 \quad \text{if } s > 0$$

or, put in other terms, the equilibrium is an attractor for $s < 0$ and a repeller for $s > 0$. Because the determinant of $\mathbf{A}$ is positive, the discriminant of the quadratic equation that determines the eigenvalues will be negative when trace $\mathbf{A}$ is close to zero. Hence there is a range of s values, say $|s| < \delta$, for which the eigenvalues are complex. The fact that the determinant is nonvanishing also guarantees that the imaginary part of the eigenvalues is nonzero at $s = 0$ even though the real part is zero there.

Our previous analysis of the fishery problem revealed an up-and-down oscillation in the values of the fish stock x as the harvesting parameter E slowly varied (Figure 7.3). This suggests that we might expect a bifurcation to an attracting cycle to emerge when s exceeds zero. Thus, even though we have no direct evidence that the equilibrium is an attractor for $s = 0$, it is quite plausible that this is in fact the case.

To complete our analysis we need to establish that the derivative with respect to s of the real part of the eigenvalues, namely, the derivative of (trace $\mathbf{A}$)/2, is positive at $s = 0$. Now $x_\star = x_m - s$ and, therefore, by utilizing the chain rule of differentiation one computes this derivative to be

$$\frac{dg'(x_\star)/ds - g(x_\star)/x_\star}{2} = \frac{-g''(x_\star) - g'(x_\star)/x_\star + g(x_\star)/x_\star^2}{2}$$

and, because of (7.9), the second two terms within braces cancel leaving

$$\frac{-g''(x_m)}{2}$$

at $s = 0$. By virtue of (7.10) the last term is positive and so the final condition required for Lemma 7.1 to hold is now fulfilled. This shows that a stable cyclic trajectory (a periodic attractor) in the x, E plane will be found for some range of $s > 0$ values. This requires that $x_\star = c/vp < x_m$, which is the same condition for oscillatory behavior of fishery collapse and recovery that we found in Section 7.2 using, however, a quite different bifurcation analysis in which E was a slowly varying parameter while x rapidly attained its equilibrium value relative to E.

The two treatments of the same model give qualitatively similar results although the specifics of how x and E change periodically over time will be different. What we learn from these disparate approaches is that the "boom-and-bust," collapse-and-recovery scenario is a good metaphor for actual fishery dynamics.

A numerical simulation of the nonlinear fishery equations verifies the existence of an attracting cycle whenever c/vp is less than x_m and of a point attractor otherwise (Figure 7.8).

7.6. Exercises

7.6.1. In the fishery model of Section 7.2, suppose that $c/vp > x_m$ and show that the equilibrium of Equation (7.3) is asymptotically stable by employing Lemma 6.2. This shows that if the cost-to-price ratio is high enough, it may be possible to avoid a fishery collapse.

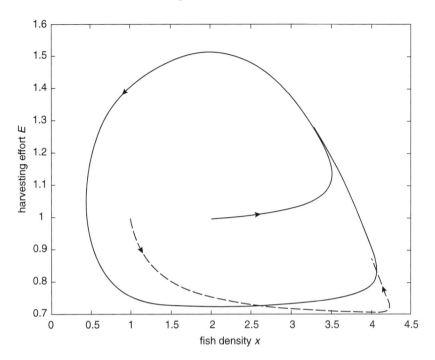

Figure 7.8. Trajectories of the fish harvesting model Equations (7.2) and (7.3) for different values of the cost to price ratio c/p using different starting values of x and E. The solid line indicates the orbit for low c/p and the dotted line is for high c/p.

7.6.2. The right side of Equation (7.3), namely, $g(x) - vEx$, is some function of x and E that can be written as $G(x, E)$. Then $G(x, E) = 0$ implicitly defines x as a function of E. In fact, the implicit function theorem of the calculus tells us that if the partial derivative $G_x(x, E) \neq 0$ at x_o, E_o, then there is some continuous function $x = h(E)$ defined in a neighborhood of E_o for which $G(h(E), E) = 0$ and $x_o = h(E_o)$. Explain why this implies that at a bifurcation point, where the equilibrium value of x jumps suddenly downward or upward (fishery collapse or recovery), is where the pair of equations $G = 0$ and $G_x = 0$ must be simultaneously satisfied.

7.6.3. A generic predator-prey model (a variant of the celebrated Lotka-Volterra equations) is defined by the following equation pair in which x denotes prey density and y the predator density:

$$x' = rx(1 - x) - \frac{bxy}{1 + x}$$

$$y' = sy\left(1 - \frac{y}{x}\right)$$

(7.11)

The growth rate of the prey is logistic and predation rate has a satiation term that tells us that if the prey is plentiful (large x) then predation is proportional to predator density solely, whereas if prey is scarce (x small), the predation is proportional to the number of encounters (namely, the product xy) between species. The growth rate of the predator is also logistic except that maximum sustainable population level of the predator diminishes as the prey becomes scarce.

Find the sole equilibrium of (7.11) in the positive quadrant of the x, y plane (it doesn't make sense to have negative population densities) and compute the Jacobian of the linearized equations at the equilibrium. Apply Lemma 7.1 to check for the conditions that would suggest a bifurcation from a point attractor to a cycle.

7.6.4. There is a body of evidence suggesting that the sudden demise of an algae bloom of the type described in Section 7.3 is due to viral contamination. When algae cells reach high concentrations during an episode of profligate growth, it becomes easier for a virus to infect the cells by penetrating the membrane wall where it replicates itself until the cell bursts (cell lysis) releasing a multitude of new viruses that can then infect healthy cells in the surrounding medium. If the cell concentration is sparse, it is more difficult to a virus to find a nearby host cell. A simple model of this phenomenon follows considerations quite similar to the epidemic model of Section 6.4. The algae cells are divided into classes of susceptible and infected, with concentrations indicated by S and I, whereas the viruses have concentrations P and these quantities interact as in the schematic of Figure 7.9. The irreversibly damaged infected cells contribute to cell density even though they do not reproduce and collectively the cells have a maximum population density K according to a logistic law with a constant per capita rate r. The healthy cells become infected at a rate proportional to the number of encounters between these cells and the viruses and the viruses are depleted at the same rate. The infected cells are lysed at a constant per capita rate α (the reciprocal of this rate is the latency period, as in the model of Section 6.4), while the viruses reproduce at a rate that is some multiple b of the depletion rate of infected cells (b is generally of the order of tens to hundreds). The viruses die off at some constant rate μ that depends on a number of hostile factors in the waters. Putting all this together gives rise to the following differential

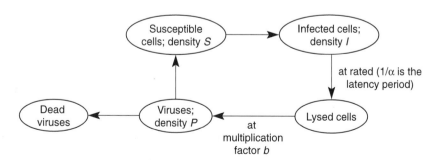

Figure 7.9. Schematic of a model for the viral contamination of algal cells.

equations:

$$S' = rS\left(1 - \frac{S+I}{K}\right) - kSP$$

$$I' = kSP - \alpha I$$

$$P' = -kSP - \mu P + b\alpha I$$

One equilibrium to this system of equations is the disease-free situation in which $(S, I, P) = (K, 0, 0)$. Show that when the multiplication factor b is large enough the equilibrium becomes unstable. In fact, solutions tend to a nontrivial equilibrium at which the viral contamination becomes endemic. Even further increases in b result in a Hopf bifurcation leading to oscillatory behavior; this consequence of the model equations would be difficult to observe in an actual ocean environment.

7.6.5. Show that the conditions of Lemma 7.1 are equivalent to finding that at $s = 0$ the determinant of $\mathbf{A}(s)$ is positive, the trace of $\mathbf{A}(s)$ is zero, and the derivative of the trace with respect to s is positive.

7.7. Further Readings

A good source for the exploitation of renewable resources and, in particular, fish harvesting models, is the book [23]. A highly recommended discussion of differential equation modeling is in a little booklet [48], where the interaction between an insect, the spruce budworm, and the forests it defoliates, is treated in a most instructive manner. At the same time, it provides an introduction to catastrophe models, a topic we alluded to in Section 7.2

The notion that a renewable resource such as a fishery, forest, or grazing lands, which are open to exploitation to all without restriction, will ultimately be devastated in the stampede to exploit it in one's own interest, without regard to the common good, has been called "the tragedy of the commons" in an elegant article by Garrett Hardin [29].

The fishery problem continues to resurface (for example, see "Northwest fishermen catch everything, and that's a problem," *New York Times*, Nov. 13, 1988) and so models like the one discussed in Section 7.2 remain appealing even if they are admittedly too simple.

A discussion of the algal bloom problem as it relates to brown tides can be found in the paper [7]. The Hopf bifurcation theorem of Section 7.4 is discussed further, with additional examples, in the author's book [10].

The model in Exercise 7.6.4 is based on unpublished notes of A. Okubo and elaborated on in [12] with similar considerations in [9]. Evidence for the viral contamination of algae was obtained by E. Cosper and others ("Natural Virus Said to Check Brown Tides," *New York Times*, Nov. 4, 1994).

Red Tides and What Ever Happened to the Red Squirrel?

8.1. Background

On a day in late spring one begins to notice patches of discolored water in the bay, dark brown like mud. A few days later the patches have spread and after several more weeks tangled masses of seaweed liter the beaches and the bay fishermen complain about the poor harvest of shellfish. Elsewhere large patches of red appear off the coast, further out at sea, and thousands of dead fish pile up on the coastline.

This may appear to be the opening scene of a sci-fi movie but in fact it is the stuff of a commonplace occurrence worldwide, year after year. The red, brown, and green splotches are dense accumulations of microscopic algae that in one way or other are toxic to the marine life around them. These algae "blooms" find conditions in the lakes, estuaries, and off-shore waters propitious for unimpeded and rapid growth until their large numbers begin to act as a brake and millions die off, leaving in their wake oxygen starved waters as the cells decay.

Although something of the temporal dynamics of these blooms was discussed in the previous chapter, our concern here will only be with the spatial distribution of these cell masses. The interplay between reproduction of algae by cell division and their dispersion in the turbulent waters sets a lower bound on the size of these plankton patches, a topic that will be taken up in Section 8.3.

A few "killer bees," a species originally imported from Africa, were accidently released in Brazil in recent times and they began to colonize portions of South and Central America by migrating in swarms and, by now, they have reached our border. In the last century, some captive gypsy moth larvae escaped in Massachusetts and spread over the Northeastern States. The proliferation of both these insects over wide-ranging areas has been a cause of distress to many.

A similar event took place around the turn of the century when the American gray squirrel, *Sciurus carolinensis*, was released into the English countryside,

where it spread to colonize much of the habitat formerly occupied by the indigenous red squirrel, *Sciurus vulgaris*. The influx of the gray squirrel has coincided
with the decline and disappearance of its red relative and although there may
be several reasons for this decline, such as disease and environmental changes,
a most plausible hypothesis is that it has been out-competed by the larger gray
squirrel, which is also known to be a more prolific breeder. It is likely that a substantial overlap in the niches of these two species has favored the advance of the
grays over the reds. The interplay between competition and dispersion gives rise
to interesting wavelike patterns of squirrel densities in space that are in rough
agreement with those actually seen in England and Wales. This idea is explored in
Section 8.5

Another example of wavelike spatial dispersion occurs in the spread of some
diseases. The epidemic model that was considered in Chapter 6 can be extended
to include the effect of a pathogen moving through a population of susceptibles.
This is carried out in Exercise 8.6.5.

The mathematical apparatus necessary for these studies in diffusion is developed
in the next section, where a certain partial differential equation is derived that
connects dispersion in space to changes that take place over time.

8.2. Diffusion

When a bunch of particles collide, each is scattered at random, first in one direction
and then another, in short back-and-forth excursions known as molecular diffusion.
In addition to this ricocheting is the dispersal that takes place at large spatial scales
due to the fact that the particles may be in a medium that is itself in haphazard
motion. Algae cells in the ocean, for example, are subject to the vagaries of wind
and tide. The cells are caught up in swirling eddies of a churning sea that tosses them
to and fro. Although it may be a leap of the imagination to go from microscopic
cells to animals, it is sometimes also appropriate to consider the dispersal of small
mammals, such as squirrels, as they migrate across fields and wooded glens, to
be another form of diffusion especially when, to all appearances, their motion is
erratic and essentially unpredictable.

A simple model of diffusion on the one-dimensional x axis is a random walk.
Let a particle make a series of short movements of length Δx, each of duration Δt,
beginning at the origin. A move one step to the right takes place with probability
p and one step to the left with probability q, $p + q = 1$. Each move is assumed
independent of the others so we have a Markov chain of the type considered in
Chapter 1, in which the one-step transition probabilities are $p_{i,i+1} = p$, $p_{i,i-1} = q$,
and $p_{i,j} = 0$ for all i and $j = i$. The only difference in the present context is that
the number of states is now infinite.

After n moves the particle is at some position $x = r\Delta x$, where r is one of
the integers, positive or negative. Let $u_{r,n}$ be the probability of being at x at time
$t = n\Delta t$. The particle can reach x in one step from either the right or the left, and

by conditioning on these disjoint events (see the appendix), it follows that

$$u_{r,n+1} = pu_{r-1,n} + qu_{r+1,n} \qquad (8.1)$$

with $u_{0,0} = 1$ and $u_{r,0} = 0$ for $r = 0$.

We now assume that there is a continuous distribution of particles (cells or animals, in our case) whose concentration at position x and time t is some function $u(x, t)$. Of course, a bunch of discrete particles can never be smeared continuously across the axis but, for large numbers of particles, this is an acceptable fiction. If the integral of the density function u is normalized to unity on the x axis, then the probability of finding a particle between x and $x + \Delta x$ at time t is

$$\int_x^{x+\Delta x} u(s, t)\,ds \qquad (8.2)$$

The function u is taken to be smooth in the sense of possessing continuous partial derivatives of the second order in x and of the first order in t. Then, by Taylor's theorem in the calculus for functions of two variables, we obtain

$$u(x, t + \Delta t) = u(x, t) + u_t(x, t)\Delta t + o(\Delta t)$$

where $o(\Delta t)$ denotes terms of the order Δt^2 and higher, and u_t designates the partial derivative of u with respect to t. In a similar fashion, we also get

$$u(x + \Delta x, t) = u(x, t) + u_x(x, t)\Delta x + \frac{u_{xx}(x, t)\Delta x^2}{2} + o(\Delta x^2)$$

in which the subscripts indicate first and second partial derivatives with respect to x and the "o" term consists of quantities in x^3 and higher.

To put relation (8.1) to work for us, we employ a heuristic argument that is actually valid only in the limit as the step sizes go to zero. Suppose that Δt is very small. Then it is unlikely that a particle to the right or left of x at a distance Δx will change its direction during this brief interval and so (8.1) can be written approximately as

$$u(x, t + \Delta t) = pu(x - \Delta x, t) + qu(x + \Delta x, t) \qquad (8.3)$$

When $p > q$, there is a net drift to the right, as the particles move about at random. Therefore we take p and q to equal $1/2$ to represent a walk without directional bias.

Substituting the Taylor expansion terms into (8.3) and dividing by Δt we easily see that a number of terms cancel and what remains is

$$u_t = \frac{\Delta x^2}{\Delta t}u_{xx} + \frac{o(\Delta t)}{\Delta t} + \frac{o(\Delta x^2)}{\Delta t}$$

Now pass to the limit as Δt and Δx go to zero, in such a way that the ratio $\Delta x^2/\Delta t$ remains equal to a constant, call it $2D$. D is known as the *diffusion coefficient* and it measures the rate of spread of particles in units of distance squared per unit time. A large D indicates a higher degree of dispersion than one with a smaller value.

Keeping in mind that $o(\Delta t)/\Delta t$ goes to zero as $\Delta t \to o$, we see that in the limit, the preceding expression becomes

$$u_t = Du_{xx} \tag{8.4}$$

which is the *diffusion equation*. In the study of heat flow, where particles disperse by collisions, (8.4) is called the heat equation.

We now offer a second derivation of (8.4) using a new hypothesis about the mechanism of dispersion. Letting $q(x, t)$ be the net flow rate of particles in units of distance per unit time, we stipulate that the flow is proportional to the rate at which the concentration of particles varies in space, and that this takes place in the direction of decreasing concentration. What this says is that diffusion is more prominent in those regions where the concentration of the particles changes most rapidly, and that the ensuing motion tends to level the differences in density by causing the flow to move from high to low concentrations. Mathematically this is expressed as

$$q = -cu_x \tag{8.5}$$

for some constant of proportionality c. This equation is sometimes called *Fick's law of diffusion* and, in the context of heat flow, as *Newton's law of cooling*. It is an empirical relation obtained from experimental observations.

The quantity $u(x, t)$ is the density of particles measured as mass per unit distance. The total mass within an interval of length Δx is given by (8.2) and so the rate of change in mass within Δx at time t is the derivative of (8.2). This can be taken inside the integral sign to give

$$\int_x^{x+\Delta x} u_t(s, t)\, ds$$

Now let $w(x, t)$ denote the net rate at which particles are added to the interval from external sources per unit time. The function w measures the net difference between sources and sinks and is assumed to be known. If w is positive, there is a net inflow of particles, with an outflow when it is negative. For example, algae reproduce by cell division and die and a simple expression for $w(x, t)$, in this case, is $\gamma u(x, t)$, where the constant γ is the net per capita difference between birth and death rates and u represents the concentration of cells. This assumes that the population density is not too large (see, in this regard, the discussion in Section 6.2). For ease of language, the function w will be generically called a source term even when it is negative.

At this point, one invokes the idea of *conservation of mass*, which stipulates that the total mass remains constant, accounting for all losses and gains. With this precept in mind one requires that the rate of change in mass within Δx to equal the changes due to w within the same interval plus the difference in flow into and out of the boundary. The net flow into the interval at x is $q(x, t)$ and is $q(x + \Delta x, t)$ at the other end and so, combining all the pieces of the flow, we find that

$$\int_{x}^{x+\Delta x} u_t(s, t)\, ds = q(x, t) - q(x + \Delta x, t) + \int_{x}^{x+\Delta x} w(s, t)\, ds$$

By Taylor's theorem, again, this time for functions of a single variable, the terms in q can be written as $-q_x(x, t)\Delta x + o(\Delta x)$, where, as before, $o(\Delta x)$ are quantities that vanish faster than Δx. The same Taylor's theorem enables us to rewrite the integrals appearing in the above expression as $u_t(x, t)\Delta x + o(\Delta x)$ and $w(x, t)\Delta x + o(\Delta x)$, respectively. Dividing through by Δx one gets

$$u_t = -q_x + w + \frac{o(\Delta x)}{\Delta x}$$

and now letting Δx go to zero, one finds that

$$u_t = -q_x + w \tag{8.6}$$

Using the fact that q is related to u through (8.5) we finally obtain the following partial differential equation in which the constant c is equated to the diffusivity D.

$$u_t = D u_{xx} + w \tag{8.7}$$

When there are no external sources or sinks this reduces to the diffusion Equation (8.4). This shows that this derivation of the equation using a conservation of mass argument is consistent, at least, with the random walk hypothesis.

A different form of dispersion occurs when a substance of concentration u is constrained to move in a specific direction along the x axis as a result of being immersed in a medium that is itself in motion at some velocity v. This directional flow, called *advection*, is the opposite to random dispersion. An example would be the placement of a red dye, say, into a river that is moving downstream in a specific direction. The meandering of the river as it makes its way to the sea is a curve whose arc length is measured along the scalar x axis. Parenthetically, the dye is a metaphor for a pollutant that is mixed in coastal waters that move seaward for awhile, and then reverse themselves, as a result of tidal motion.

In the case of advection the flow rate q may be expressed as $q = vu$, in units of distance per unit time multiplied by mass per unit distance, namely, mass per unit time, where the velocity v is itself a function of x and t. Inserting this expression into conservation of mass Equation (8.6), one gets

$$u_t = -(vu)_x + w \tag{8.8}$$

which is known as the *advective equation*.

It is useful to show what Equation (8.7) looks like in the context of two spatial dimensions x and y because this is more appropriate to algal motion that takes place on the surface of the sea or to the terrestrial migrations of insects and animals along grasslands and woods.

The arguments are really a repetition of what was done before except that motion takes place in a plane instead of along a line, with the continuous distribution $u(x, t)$ replaced by $u(x, y, t)$.

Consider a region R in the plane enclosed by a smooth curve C whose coordinates are parametrized by $\mathbf{r}(s) = (x(s), y(s))$, where s denotes arc length. The tangent to the curve at s is therefore given by the derivative $\mathbf{r}'(s) = (x'(s), y'(s))$. It is not difficult to see that the row vector $\mathbf{r}'$ is orthogonal to the vector $\mathbf{n}$ defined by $(y(s), -x(s))$, for all s, and we say that $\mathbf{n}$ is normal to the curve C. The vector $\mathbf{n}$ is perpendicular to C and points out of R.

The total mass of the particles inside R is given by the double integral of the concentration u over R and its time rate of change is, as before, the derivative with respect to t, taken inside the integral sign:

$$\iint_R u_t(x, y, t)\, dx\, dy \tag{8.9}$$

The conservation of mass argument stipulates that the rate of change in mass of particles in R, as given by (8.9), must equal the net flow of particles across the boundary of R plus any additional sources and sinks within R. As before, the rate of flow of sources and sinks is some known source function $w(x, y, t)$ measured as concentration per unit time.

Let $\mathbf{q}(x, y, t)$ be the flow of particles in units of mass per unit time. This is a vector with components q_i, $i = 1, 2$. Because $\mathbf{n}$ is an outward normal to C, the inner product of $\mathbf{q}$ and $\mathbf{n}$, denoted by $\mathbf{q} \cdot \mathbf{n}$, is the component of $\mathbf{q}$ in the outward direction, and the integral of this quantity over the boundary curve C is the net outward flow of particles from R:

$$-\int_C \mathbf{q} \cdot \mathbf{n}\, ds \tag{8.10}$$

The minus sign indicates that an outflow decreases the mass within R. Now we invoke the same Fick's law that was used in the one-dimensional case by stipulating that the two components q_1 and q_2 of the vector $\mathbf{q}$ are proportional to the derivatives of u in the x and y directions: $q_1 = -cu_x$ and $q_2 = -cu_y$. In vector notation this is $\mathbf{q} = -c\nabla u$, where ∇u denotes the gradient vector of $\mathbf{u}$ with components u_x and u_y.

The line integral (8.10) can now be written as

$$\int_C \mathbf{q} \cdot \mathbf{n} \, ds = -\int_C (q_1 y' - q_2 x') \, ds = -c \int_C (u_y x' - u_x y') \, ds \quad (8.11)$$

and, by Green's theorem in the calculus, the line integral on the right of (8.11) equals the double integral

$$c \int_R \int (u_{xx} + u_{yy}) \, dx \, dy \quad (8.12)$$

Combining (8.12) with (8.9) and remembering the source function w, conservation of mass tells us that

$$\int_R \int (u_t - c\nabla^2 u - w) \, dx \, dy = 0$$

where $\nabla^2 u = u_{xx} + u_{yy}$. Since R was chosen to be any set in the plane that is bounded by a smooth curve, the integrand is itself zero for all x and y. Otherwise, if we suppose that the integrand is positive, say, at some point in the plane, one can find a small disk I about that point in which the integrand remains positive. This is true by virtue of the fact that $u_t - C\nabla^2 u - w$ is a continuous function and so cannot change its sign abruptly. It follows that the double integral of this quantity over I is also positive, which is a contradiction. The two-dimensional diffusion equation is therefore

$$u_t - D\nabla^2 u - w = 0 \quad (8.13)$$

in which we write c as D to remind us of the diffusion coefficient.

In problems with circular symmetry, such as a two-dimensional flow of particles that is isotropic (meaning that the flow depends on radial distance but not on the angle θ chosen), Equation (8.13) can be written differently. Make a change of variables by letting $x = r \cos \theta$ and $y = r \sin \theta$ and define a function v by $v(r, t) = u(r \cos \theta, r \sin \theta, t)$, independent of θ. Using the chain rule of differentiation for two variables one can show (Exercise 8.6.2) that

$$u_{xx} + u_{yy} = \frac{v_r}{r} + v_{rr}$$

and therefore, with $W(r, t) = \mathbf{w}(x, y, t)$, the diffusion equation becomes

$$v_t = \frac{Dv_r}{r} + Dv_{rr} + W \tag{8.14}$$

8.3. Algal Patches

Several species of algae form red patches in the sea when their concentrations increase. Outbreaks of these "red tides" have been observed in many parts of the globe from the coastal waters of Japan to the Gulf of Mexico to the Adriatic Sea, and it is generally accepted that these patches are bodies of water that are especially favorable to the growth of the small plankton organisms. Conditions that contribute to this growth are an accumulation of stimulatory nutrients and levels of temperature and salinity that predispose the organisms to reproduce. These changes are brought about by a variety of causes such a storm episode that alters the composition of the water body or an upwelling of nutrients from deep layers of the sea. Tidal mixing and wind shears eventually break up the patches, but while they last the algae multiply rapidly. Outside the patches conditions are less propitious to growth and an organism that is swept outside the patch boundary by the small-scale dispersive action of wind and wave is considered lost. There is an interplay, therefore, between the aggregative process of cell division and the antiaggregative action of diffusion that carries the cells to waters that are physiologically unsuited to reproduction (Figure 8.1). Because growth takes place within the patch, it is proportional to surface area, whereas losses take place at the boundary and so are proportional to circumference. Therefore, as the area of the patch decreases, the effect of diffusion at the boundary becomes more significant, and eventually a limit is reached at which reproduction can no longer compensate for the losses. Thus there must be a minimum patch size if a bloom is to form and that is what

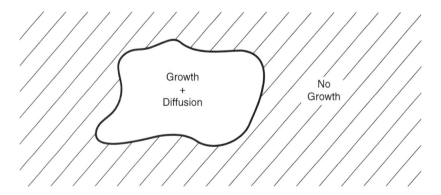

Figure 8.1. Schematic of an algal patch.

we want to show next. Diffusion at the boundary acts as an inhibitor to the growth taking place internally to the patch and it is the interplay of the two, activation and inhibition, that generates the interesting dynamics. This is a variation on a similiar theme played out in the cases of epidemics and blood clotting, which were treated in previous chapters.

Our discussion of blooms here differs in several respects from that in Section 7.3. There, the dynamics occurred over time and the dominant causative factor was the interplay of different species using a fairly sophisticated form of trophic interaction. Here, instead, the focus is on dispersion in space using a cruder form of species interaction. Although in both cases the concern is bloom initiation, the mechanisms involved are different. In actuality, it is a combination of spatial and temporal factors that need to be considered together.

After this lengthy digression, let's return to the "red tide" and assume a circular region in the sea in which the concentration of algae varies with the radial distance r from the center of the patch but not with direction. This isotropy condition reflects the belief that in the turbulent diffusion of the ocean we have little reason to bias the flow along any particular path.

Algal cell reproduction is considered to be a source term that continuously adds to the population and is expressed in simplest form as $\gamma u(x, y, t)$, where γ is a constant per capita net growth rate and u is the concentration of algae at position x, y on the ocean surface at time t, so that ru is the growth rate in density. A more meticulous model of growth would be logistic, taking into account the inhibiting effect of overcrowding but, because there are continuous losses due to diffusion, the population is unlikely to reach a level that is self-limiting. In this regard, we suggest that reference be made to the discussion in Section 6.2.

We adopt the diffusion Equation (8.14) in which the source term is $W(r, t) = \gamma v(r, t)$:

$$v_t = \frac{Dv_r}{r} + Dv_{rr} + \gamma v \tag{8.15}$$

The function $v(r, t)$ equals $u(x, y, t)$ in polar coordinates. We wish to find the minimum radius L that can sustain a bloom. Outside of a circle of this radius the conditions for growth are unfavorable and so we impose the requirement that $v(L, t) = 0$.

We are not interested in actually solving (8.15). Our goal is merely to find a relationship between the parameters r and D and the patch width L. To this end we begin by reducing the partial differential equation to a pair of ordinary differential equations by guessing that the solution is of the form $v(r, t) = a(r)b(t)$ for suitable functions a and b. Whether this satisfies the equation or not remains to be seen. Make the substitution $v = ab$ into (8.15) and divide out by ab to get

$$\frac{b_t}{Db} - \frac{\gamma}{D} = \frac{a_{xx}}{a} + \frac{a_x}{ra}$$

where the subscripts denote partial derivatives with respect to the indicated variables. The left side of this relation is solely a function of t whereas the right side depends only on x. Because these are independent variables, the only way both sides can equal each other for all x and t is for each to equal some constant that we write as $-\lambda$, with $\lambda > 0$, for a reason that will be clear in a moment. This gives us two separate equations:

$$b_t - (\gamma - D\lambda)b = 0 \qquad (8.16)$$

$$ra_{xx} + a_r + \lambda ra = 0 \qquad (8.17)$$

in which the subscripts now denote ordinary derivatives. Both these equations have solutions and so the ploy of separating u in to a product of a and b is indeed workable. The first of these equations has the well-known solution

$$b(t) = b_0 e^{(\gamma - D\lambda)t} \qquad (8.18)$$

where b_0 is the value of b at the initial time $t = 0$. If we had chosen the constant as λ, instead of its negative, then (8.16) shows that b (and therefore v) would increase at a temporal rate in excess of γ, which is biologically unreasonable because the maximum growth rate that can be expected from cell divisions is γ.

The second equation, however, may be less familiar and is called *Bessel's equation*, after the German astronomer F. W. Bessel. Rather than dwell on the details we simply point out that this equation possesses a solution, the *Bessel function of the first kind of order zero*, that remains finite when $r = 0$. There is also another solution, the Bessel function of the second kind, that becomes unbounded as r approaches zero and that is discarded because this gives a physically unacceptable solution. The Bessel function we use is denoted by $J_0(\lambda^{1/2} r)$ and it has the value one when $r = 0$ and then oscillates, with its first zero at $\lambda^{1/2} r = 2.405$ (Figure 8.2). Chapter 8 of the book by Jeffrey [31] supplies the missing details on the Bessel equation and its solutions.

At this point we must model the effect of the patch boundary, where we imposed the condition $v(L, t) = 0$, for all t, to capture the notion that the waters are inhospitable to growth outside the patch. Because $b(t)$ is never zero, the product of a and b is zero at $r = L$ only if $J_0(\sqrt{\lambda}L)$ vanishes. The smallest value of L for which this is true is $L = 2.405/\lambda^{1/2}$, which shows that $\lambda = (2.405/L)^2$. Finally we see from (8.18) that if $\gamma - D\lambda$ is negative, then $u(x, t)$ tends to zero as t goes to infinity. But we are seeking conditions that guarantee a sustainable growth within the patch and so we insist that $\gamma - D\lambda$ be nonnegative. Using the value of λ already obtained, this gives the condition $\gamma - D(2.405/L)^2 \geq 0$ or, put another way,

$$L \geq 2.405\left(\frac{D}{\gamma}\right)^{1/2} \qquad (8.19)$$

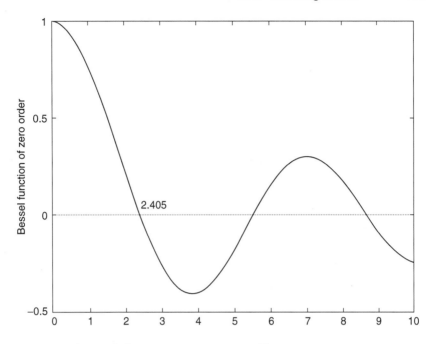

Figure 8.2. The Bessel function $J_0(\sqrt{\lambda}x)$ for λ equal to one.

The right side of this inequality is the critical patch size, the smallest width that allows the algae to provide a sufficient number of new recruits to replace those lost by dispersion. Anything smaller and the reproductive ability of the organisms is outstripped by diffusion. Note that as the per capita growth rate γ increases, the critical size is allowed to get smaller, whereas if dispersion increases, as measured by the constant D, L must also increase. Both results are consonant with what we expect.

It has been observed that actual bloom patches appear to vary in size from about 10 to 100 km in the open sea and from about 1 to 10 km in more protected coastal embayments. Allowing the growth rate γ to vary from 0.10 to 1 cell divisions per day, a figure that is about right in temperate-zone waters, and using estimates of the diffusion coefficient D that are worked out in the book by Okubo [42] list, the critical radius obtained from the model gives values of between 2 and 50 km, which is in reasonable accord with observations.

8.4. Traveling Waves

In the model to be discussed in the subsequent section, a wavelike solution is anticipated on intuitive grounds. A few comments are inserted here to explain what this means.

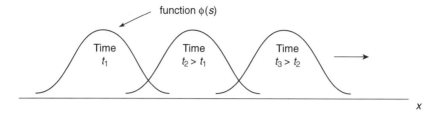

Figure 8.3. The traveling wave function $\phi(s)$ for different values of t, where $s = x - qt$.

Consider equations of the form

$$u_t = Du_{xx} + ru(1 - u)$$ (8.20)

in which there is diffusion coupled with a source term $ru(1 - u)$ that represents logistic growth. This equation is sometimes called *Fisher's equation*, after the English statistician R. A. Fisher, and is an example of a wider class of *reaction–diffusion equations*.

A *traveling wave solution* to (8.20) is of the form $u(x, t) = \phi(s)$, where $s = x - qt$ for some constant $q > 0$. Think of t as representing time and x as a one-dimensional spatial variable. Although ϕ is a function of s just think of it for a moment as a function of x with t simply a parameter. Now suppose that x is fixed and that t is allowed to increase from zero; this is equivalent to viewing the profile ϕ at successive values to the left of x. In fact, the effect is like that of seeing a movie of ϕ moving past x to the right as t increases (Figure 8.3).

Are there in fact traveling wave solutions? In Figure 8.4 an intuitive argument is presented to support the idea that one can expect such wavelike solutions to (8.20), at least in the case where the initial value of u is a step function:

$$u(x, 0) = \begin{cases} 1, & x \leq 0 \\ 0, & x > 0 \end{cases}$$

Substituting ϕ into Equation (8.20) and carrying out the indicated differentiations results in a second-order ordinary differential equation, in which prime denotes derivative with respect to the variable s:

$$-q\phi' = D\phi'' + r\phi(1 - \phi)$$ (8.21)

This equation will reappear several times in our subsequent discussion.

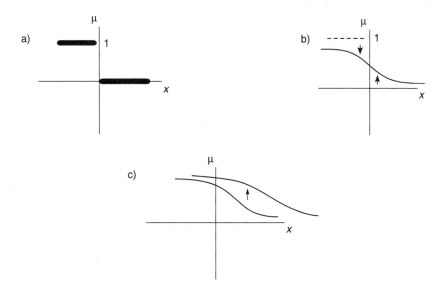

Figure 8.4. Intuitive argument for the temporal evolution of a solution of Equation (8.20) when the initial density is a step function. This is shown in a series of snapshots in space for different times. The step function is in panel (a). A bit later the step function (now dotted) has evolved to the solid line shown in panel (b). What has happened is that diffusion has robbed from the dense population and added to the sparse one. Finally, the waveform in (b), now dotted, has evolved further, panel (c). This is so because population growth is largest when μ equals 1/2 and smallest near the endpoints 0 and 1. By repeating the actions of diffusion and growth in this manner, the original profile is seen to move to the right.

8.5. The Spread of the Gray Squirrel

When the American gray squirrel was released into the English countryside about the turn of the century it was faced with a competitor, the indigenous red squirrel, that occupied the same habitats and had similar food preferences and foraging habits. Some red squirrels fell prey to a disease, possibly transmitted from its gray cousin but, perhaps more significantly, it was out-competed by the more prolific breeding and larger size of the American squirrel. As time went on, the intruder began to colonize new areas, mostly woodlands, as the original sites became over-crowded. As the invading grays advanced forward, they largely replaced the reds that had already settled there. The two species coexisted for a time as the concentration of the grays increased and the that of the reds diminished. To all appearances, then, the changes in population densities were waves moving outward in space as time evolved, two waves out of phase as the gray squirrel progressively colonized new areas and the red squirrel receded (Figure 8.5).

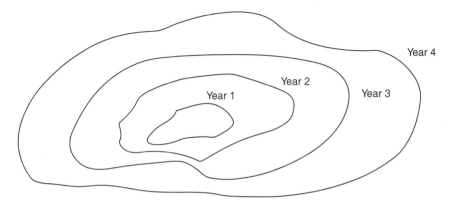

Figure 8.5. Sketch of a typical dispersion of squirrels in space, with the contours marking the extent of propagation over four successive years.

The idea now is to model the wavelike behavior of the invading squirrels by using the diffusion Equation (8.7) to represent the spatial spread of these mammals along a single axis of motion. In addition, the interaction between the two species is modeled in terms of the competition equations derived in Section 6.2. These play the role of w, the net source terms, in the diffusion equation. The restriction to a one-dimensional movement is done for reasons of mathematical expediency but does not seriously detract from the interpretative value of the model.

Let $u(x, t)$ and $v(x, t)$ be the population densities of the gray and red squirrels, respectively, which are assumed, as usual, to be twice continuously differentiable functions in x and t. Then, combining (8.7) with (6.6), we obtain a pair of equations

$$u_t = Du_{xx} + ru\left(1 - \frac{u}{K}\right) - auv$$

$$v_t = Dv_{xx} + sv\left(1 - \frac{v}{L}\right) - buv$$
(8.22)

in which the competition terms represent sources and sinks. In a paper referred to in the next section, the net per capita birth rates r and s are estimated to be 0.82 and 0.61 per year, respectively, with maximum population densities of 10 and 0.75 per hectare (which is 10^4 square meters). The diffusion coefficients are taken to be the same, at about 18 km^2 per year. The competition coefficients a and b are difficult to estimate from the available data, but it was reasonable to assume that a is less than b because the grays out-compete the reds and it was decided to pick $a = 0.5$ and $b = 1.5$.

In the interest of seeing what is at least theoretically possible, the model is simplified even more by scaling all the variables so that many of the coefficients reduce to unity. This enables us to obtain at least some insight into the model structure by analytical means. Later, the results of a numerical simulation of the

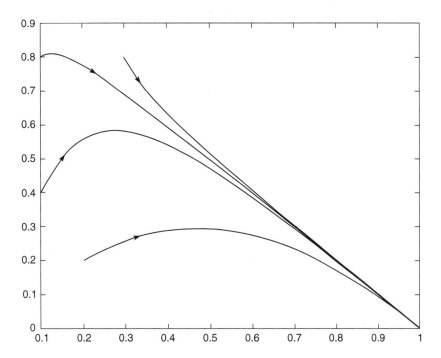

Figure 8.6. Trajectories for the model Equations (8.23) in the case of spatial independence, where $a < 1, b > 1$.

full model are compared to actual field observations. The drastically simplified equations are then

$$u_t = u_{xx} + u(l - u) - auv$$
$$v_t = v_{xx} + v(l - v) - buv \tag{8.23}$$

with $a, l, b > 1$, and $a + b = 2$.

In the absence of diffusion and spatial dependence (so that, in particular, $u_{xx} = v_{xx} = 0$), the system (8.23) consists of coupled ordinary differential equations with equilibria at $(u, v) = (0, 0)$, $(0, 1)$ and $(1, 0)$ in the nonnegative quadrant of the u, v plane. A nullcline analysis shows that the first two points are unstable but that the last equilibrium is an attractor (Figure 8.6 and Exercise 8.6.3). Figure 8.6 demonstrates that it is possible for trajectories to flow toward the point $u = 1$, $v = 0$ in which the gray squirrels have outcompeted the reds. It is another version of the paradigm that when two species are in strongly overlapping niches, one of them is inexorably driven to extinction.

We now return to the partial differential Equations (8.23) and look for traveling wave solutions of the form $u(x, t) = f(s)$ and $v(x, t) = h(s)$, where $s = x - qt$

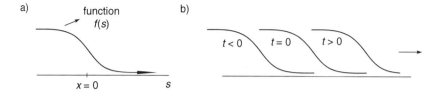

Figure 8.7. A wavelike function $f(s)$, where $s = x - qt$, in panel (a), and profiles of this function as t increases, for x and q fixed in panel (b).

for $q > 0$. As discussed in the previous section, the functions f and h move parallel to themselves in the positive x direction with constant shape as t increases. Suppose that f, for example, has the form exhibited in Figure 8.7a, which is asymptotic to 1 at $s = -\infty$ and is asymptotic to 0 at $s = +\infty$. Then, for any fixed position $x, s = x - qt$ goes to $-\infty$ as $t \to +\infty$ and a little reflection shows that $f(s)$ moves to the right at a constant speed q while maintaining its shape (Figure 8.7b). In effect $f(s)$ increases, for a given position x and speed q, only if $s = x - qt$ moves to the left, and that occurs only as t increases.

Substituting the functions f and h into (8.23) produces a pair of ordinary differential equations in the single variable s, in which all derivatives are indicated by a prime,

$$-qf' = f'' + f(1 - f) - afh$$
$$-qh' = h'' + h(1 - h) - bfh \tag{8.24}$$

Because $a + b = 2$ this set of equations can be added together to obtain a single equation in the variable $z(s) = f(s) + h(s)$, which, as is easily verified, satisfies

$$-qz' = z'' + z(1 - z) \tag{8.25}$$

This is a special case of (8.21) using different notation.

We look for traveling wave solutions to (8.24) to mimic the spatial spread of gray squirrels as they overtake the reds. In the remote past, the reds are at high concentrations and the grays are low, but as time elapses their status at any given location is reversed. To reflect this we impose the conditions that $f(s) = 0$, $h(s) = 1$ at $s = +\infty$, whereas $f(s) = 1$, $h(s) = 0$ at $s = -\infty$. This means that for any fixed x, the red squirrels are dominant as t moves toward negative infinity (the distant past) with just the opposite as t advances to the distant future. In terms of the sum $z(s)$, this translates into $z(s) = 1$ at $\pm\infty$. We will show here that this implies that $z(s)$ is identically equal to unity for all s and it follows from this that $h(s) = 1 - f(s)$. Substituting this last relation into the first of Equations (8.24) one gets

$$-qf' = f'' + f(1 - f - ah) = f'' + (1 - a) f (1 - f) \tag{8.26}$$

which is another form of (8.21), this time with the conditions $f(-\infty) = 1$, $f(+\infty) = 0$.

The second-order Equation (8.26) can be rewritten as a pair of first-order equations in a new set of variables x_1 and x_2 by letting $x_1 = f$ and $x_2 = f'$. This provides us with

$$
\begin{aligned}
x_1' &= x_2 \\
x_2' &= -qx_2 - (1-a)x_1(1-x_1)
\end{aligned}
\tag{8.27}
$$

where a is, as stipulated, less than one.

The only equilibria of (8.27) are at $(x_1, x_2) = (0, 0)$ and $(1, 0)$ and by looking at the signs of the derivatives we can reveal something of the structure of the solutions, much as was done in Chapter 6. For example, it is always true that $x_1' = x_2 > 0$ in the positive quadrant of the x_1, x_2 plane, whereas it is always negative in the lower right-hand quadrant, where $x_2 < 0$. This already tells us that trajectories move to the right (left) in the upper (lower) half planes separated by the horizontal axis where $x_2 = 0$. Because $x_1' = 0$ on this axis (the $x_1' = 0$ nullcline), the axis is necessarily crossed in a vertical direction. Therefore a trajectory that emanates from the equilibrium state $x_1 = 1$, $x_2 = 0$ can never return there since it would have to move to the left in the lower plane and return to the right along the upper plane. But the derivative x_2' is always negative in the positive quadrant whenever $x < 1$ and so, not only do the trajectories move to the right there, but they also slope downward. This precludes a closed trajectory that begins and ends at the equilibrium state other, of course, than the equilibrium state itself [namely, a constant solution to (8.27)].

We should be clear here that any solution that emanates from or tends toward an equilibrium solution can never actually reach this state because this would mean that two distinct solutions begin at the same point in the plane, which is a violation of the uniqueness principle for differential equations (see Section 6.2). Therefore, although it is possible for a trajectory to approach the equilibrium state $x_1 = 1$, $x_2 = 0$ arbitrarily close as s tends to infinity or, reversing the direction of the flow, as s tends to negative infinity, it cannot loop back to this point. If you think of s as "time" this idea is easier to grasp.

After this digression we return to Equation (8.25) which, after all, is a special case of (8.26) in which $a = 0$, and confirm that indeed $z(s)$ must be constant in s since it represents a solution that begins and ends at the equilibrium where $z \equiv x_1 = 1$, $z' \equiv x_2 = 0$.

As for Equation (8.27), the argument just given supports the possibility of a solution that emanates from the unstable equilibrium at $x_1 = 1$, $x_2 = 0$ and moves to the stable point at the origin. A formal argument can be given to verify that this is in fact true, and a numerical solution of Equations (8.27) exhibits what appears to be just such a trajectory (Figure 8.8). This provides a traveling wave solution to the first equation of (8.24) with the desired properties that $f = 1$ at $s = -\infty$ and $f = 0$ at $s = +\infty$. Moreover, because the solution lies in the lower half plane, it

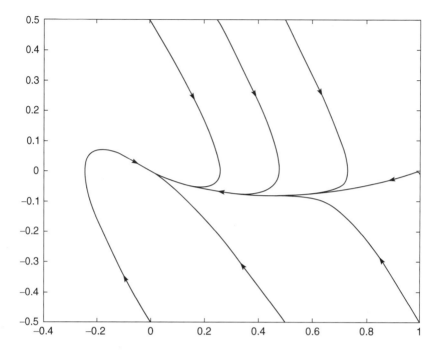

Figure 8.8. Trajectories of the model Equations (8.26) in which we see a single orbit that links two equilibria from $f = 1$ to $f = 0$, where f' is zero. This corresponds to a traveling wave solution to the first equation in (8.24).

means that $f' = x_2$ is negative and, therefore, that f is monotone decreasing, as required.

We close our discussion by estimating the minimum wave speed q and then comparing it to the empirically observed spread of the gray squirrel.

Imagine that a line L passes through the origin of the x_1, x_2 plane with negative slope $-\eta$, $\eta > 0$. If we can show that all trajectories cross this line inwardly (in an upward direction) then any solution that moves from $f = 1$ to $f = 0$ in the f, f' plane (namely, the x_1, x_2 plane) would never be able reach the origin through negative values of f because it would be confined to the region made up of the horizontal line $x_2 = 0$, the vertical axis defined by $x_1 = 1$, and the line L. The trajectory emanating from $f = 1$, $f' = 0$ never gets to be negative. This is important since a negative f implies a negative population density, which is biologically unfeasible. The existence of L is linked to the value of q as will now be seen.

The line L has coordinates $(x_1, x_2) = (x_1, -\eta x_1)$ and the normal to L is the column vector $\mathbf{n}$ that we write in row form as $(\eta, 1)$. The inner product between $\mathbf{n}$ and $(1, -\eta)$ is zero and so $\mathbf{n}$ is perpendicular to L and points up and to the right of the line. The solution to (8.27) is the vector $\mathbf{x}(s)$ with components $x_1(s)$ and the derivative $\mathbf{x}'(s)$ has components $x_i'(s)$; $\mathbf{x}'$ is tangent to the trajectory and points in the direction of the flow. Recalling that the inner product of two vectors is a measure of the cosine of the angle between them, it follows that any trajectory that crosses L does so in an inward direction, whenever the inner product between $\mathbf{n}$ and $\mathbf{x}'$ is positive, since they differ in direction by an angle that is less than $\pi/2$. What needs to be demonstrated, then, is that the inner product between $(\eta, 1)$ and the column vector with components x_2, $-qx_2 - (1 - a)x_1(1 - x_1)$ is positive along L. Using the fact that $x_2 = -\eta x_1$ along L it means that $-\eta^2 x_1 + q\eta x_1 - (1 - a)x_1(1 - x_1)$ is positive or, factoring out x_1, the requirement is that

$$\eta^2 - q\eta + (1 - a)(1 - x_1) < \eta^2 - q\eta + (1 - a) < 0$$

since this gives the positivity we are after (note: $0 \le x_1 \le 1$).

The quadratic $\eta^2 - q\eta + (1 - a)$ is positive when η is zero and when η is sufficiently large. Therefore the quadratic can be negative only if it has two positive roots. But the roots of $\eta^2 - q\eta + (1 - a)$ are readily computed to be $\eta = q/2 \pm (q^2 - 4(1 - a))^{1/2}/2$ and so are real and positive only if $q^2 \ge 4(1 - a)$. It follows that there is a critical minimum speed q defined by

$$q \ge 2(1 - a)^{1/2} \tag{8.28}$$

if there is to be a feasible travelling wave solution.

Observe that if $f(s) = z(s) - h(s)$ is now inserted into the second of Equations (8.24) another form of Equation (8.21) is obtained, this time for the function h:

$$-qh' = h'' - (b - 1)h(1 - h) \tag{8.29}$$

with $b > 1$. The conditions on h are $h(-\infty) = 0$ and $h(+\infty) = 1$. An analysis that is similar to the one just concluded for the Equations (8.27) also applies to (8.29) (Exercise 8.6.4) showing that there is a minimum speed q given by $q \ge 2(b-1)^{1/2}$. But $a + b = 2$ and so the minimum speeds for f and h are the same. The gray and red squirrels advance and recede at the same rate.

In a study done by Okubo and others [43] the minimum speed is estimated from observations to be about 7.7 km per year. This how fast the gray squirrel encroaches on new terrain occupied by the red squirrels. Using estimates of growth rates and diffusion that were mentioned earlier, the model predicts a wave speed of 7.66 km per year. Not bad!

8.6. Exercises

8.6.1. Consider a random walk on the integers $0, 1, \ldots, N$ with one-step transition probabilities $p_{1,i+1} = p$, $p_{1,i-1} = q$, $p + q = 1$, $p_{0,0} = p_{N,N} = 1$, and $p_{i,j} = 0$ for all other i, j. In the terminology of Chapter 1, this is a Markov Chain on $N + 1$ states, in which 0 and N are absorbing states. Let b_1 be the probability of ever reaching state N if the process begins in state i (this is $b_{1,N}$ in the notation of Chapter 1).

State N can be reached from i by first taking one step to the right or left. Conditioning on these disjoint events gives

$$b_i = pb_{i+1} + qb_{i-1} \qquad (8.30)$$

The reasoning here is similar to that employed in deriving relation (8.1) in the text. Now b_i is identical to $(p+q)b_i$ and so (8.30) can be rewritten as

$$p(b_{i+1} - b_i) = q(b_1 - b_{i-1}) \qquad (8.31)$$

Using the fact that $b_0 = 0$, together with (8.31), show that

$$b_i = b_1 \sum_{k=0}^{i-1} \left(\frac{q}{p}\right)^k$$

for $i = 1, 2, \ldots, N-1$.

Now recall the following identity for the partial sum of a geometric series of a scalar a:

$$\sum_{k=0}^{r} a^k = \frac{1 - a^{r+1}}{1 - a}$$

for any integer $r > 0$. From this, and the fact that $b_N = 1$, show that

$$b_i = \begin{cases} (1 - (q/p)^i)/(1 - (q/p)^N) & p \neq q \\ i/N & p = q \end{cases}$$

Compare this result with Exercise 1.5.4 for the case $N = 4$, which was obtained earlier using a different argument.

8.6.2.   We are given a smooth function $u(x, y, t)$. Make a change of variables $x = r \cos \theta$, $y = r \sin \theta$. If u is independent of θ, then $u(x, y, t) = v(r, t)$ for some function v. Using the chain rule of differentiation, and the notation of Section 8.2, establish that $u_{xx} + u_{yy} = v_r/r + v_{rr}$.

8.6.3.   Assume that u and v are independent of x in Equation (8.23) to obtain a pair of ordinary differential equations in the independent variable t. These have three equilibria in the nonnegative quadrant of the u, v plane. Using either the nullcline or the eigenvalue (Lemma 6.2) approaches developed in Section 6.3, show that two of these are unstable but that the remaining one is an attractor.

8.6.4.   Find the minimum wave speed of the traveling wave solution to Equation (8.29) using an argument similar to that employed for Equation (8.27).

8.6.5. In its simplest form the epidemic model of Chapter 6 consists of two equations, one for the fraction S of susceptibles and the other for the fraction I of infectives [Equation (6.14) and Exercise 6.7.6]. A third group, the fraction R of recovered individuals is obtained from the relation $S + I + R = 1$.

This model can be extended by allowing for the spatial spread of a disease. A specific case is the propagation of rabies by infected foxes that wander at random through a population of healthy but susceptible foxes. The uninfected animals have territorial instincts that keep them more or less fixed in space but the sick animals appear to lose their sense of territoriality and they move about. The disease itself is transmitted through the saliva of rabid foxes.

Equations (6.14) are modified by letting S and I be functions of a one-dimensional space axis x as well as time t, with diffusion confined to the rabies infected foxes:

$$S_t = -bSI$$
$$I_t = bSI - cI + DI_{xx}$$

(8.32)

where D is the diffusion coefficient that measures the rate of dispersion of the sick animals. The contact rate is b, as before, but c now represents the removal rate of sick foxes by death, instead of a recovery rate, since rabies is invariably fatal to the infected animals.

We seek traveling wave solutions to (8.32) in the form $s(x, t) = f(z)$ and $I(x, t) = h(z)$, where $z = x - qt$ and $q > 0$ is the wave speed. We suggest you refer to the discussion in Section 8.4 on traveling waves noting, however, that s has been replaced by the variable z to avoid confusion with S. In the distant past, the population of susceptibles at any position x has the same constant density $S = 1$ (no infectives have turned up yet and there are no deaths of sick animals). A pulse wave of rabid foxes moves through space and decreases the percentage of healthy foxes to some value $S_m > 0$. The pulse is zero in the remote past and in the distant future (see Figure 8.9). Formulate the proper conditions on f and h at z at plus or minus infinity to reflect these hypotheses.

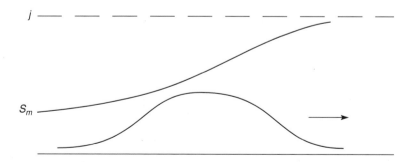

Figure 8.9. A pulse wave of infectives traveling through a territory in which there is a fixed number of susceptibles.

Substitute f and h into (8.32) and show that

$$-qf' = -bfh$$
$$-qh' = bfh + ch + Dh''$$

(8.33)

At the leading edge of the moving pulse wave, as the infectives reach the hitherto uninfected foxes, S is close to one in value so the second of Equations (8.33) can be approximated at the leading edge by

$$h'' + qh' + (b - c)h = 0$$

where D is scaled to one for notational convenience. This is a linear second-order differential equation and, using standard solution techniques for such equations (see [31]) show that the solution h is proportional to the exponential of

$$\frac{(q \pm (q^2 - 4(b - c))^{1/2})z}{2}$$

If $q^2 < 4(b - c)$ the solution is complex and $h(z)$ will oscillate about the z axis, taking on positive and negative values alternatively. But we must insist that h remain positive in order that it represent a biologically feasible population density. Therefore,

$$q \geq 2(b - c)^{1/2}$$

is the minimum wave speed. Note that c/b must be less than one for a positive speed, which is the same threshold condition that was arrived at in Exercise 6.7.6. Even though this reasoning is based on an approximation, a more careful argument along the lines of that given in Section 8.5 would allow us to reach the same conclusion.

8.7. Further Readings

Recurrent toxic blooms are simply one manifestation of how coastal waters are being stressed by the impact of human activity, and the perception that our waters are deteriorating is cause for increasing anxiety ("Brown Tide Hits Peconic and Baymen Fear Worst," *New York Times*, June 30, 1991, and "Algae Bloom Outbreaks Seen in Ocean: Spread of 'Red Tides' Could Harm Fisheries," *Washington Post*, October 31, 1991). An introduction to red tides, in general, is the *Scientific American* article [1].

The plankton patch model follows an early paper on the subject [47], whereas the squirrel migration model is based on a more recent publication by Okubo and others [43]. An overview of diffusion models of algae, insects, and animals is contained in the excellent book by Okubo [42], and we also recommend the substantial treatise [40] that includes diffusion models in the study of epidemics, as well as a derivation of the diffusion equation.

The spread of organisms that prove to be a menace as they colonize a new territory is also illustrated by the study of immigrant snails that decimate an indigenous species of snails, as engagingly told by S. J. Gould in [28].

Submarines and Trawlers

9.1. Background

In the gripping movie *Run Silent, Run Deep*, Clark Gable plays the role of a Second World War submarine commander who is obsessed with penetrating enemy defenses without being detected. In one episode, the sub lies motionless in the depths of the ocean, engines shut down, to prevent the enemy from picking up any sounds that would reveal the position of the oblong vessel. The captain of the menacing hulk on the surface is intent on destroying the U-boat and both commanders are engaged in a deadly game of evasion and pursuit.

Another encounter, with economic consequences this time, but equally in earnest, is the competition between two individuals or two firms for the right to exploit a common resource. To put this in perspective we return to the fishery problem of Chapter 7 in which access to a fishery is open to only two participants, both of whom want to maximize their own revenues but are hindered by the presence of the other. This too can be thought of as a game.

The circle of ideas concerning harvesting of resources and competition that were begun earlier in the book are brought together in this chapter and are used to illustrate some elementary techniques in optimization that require the use of calculus.

In the next section a useful lemma from the calculus of variations is established that is then employed in Section 9.3 to discuss optimal tactics for a submarine and its searcher. The fishery problem is treated next by introducing an optimal strategy between competitors that has a different twist.

9.2. A Variational Lemma

If $f(x)$ is a real-valued function of a single variable x and if f has a maximum at some point x_0 in an open interval $a < x < b$, then, assuming that f is differentiable in this interval, we know from the calculus that $f'(x_0) = 0$. However, if the

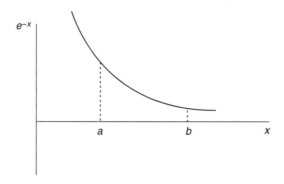

Figure 9.1. The function $f(x) = e^{-x}$ has a maximum at $x = a$, where $f'(a) \neq 0$.

maximum occurs on the boundary of the interval, at endpoint a say, then it is no longer necessarily true that the derivative vanishes there (Figure 9.1). In fact, $f(a + h) - f(a) < 0$ for all small enough positive h, and so if one divides this difference by h and then takes the one-sided limit $f'_+(a)$ as $h \to 0$ through positive values, the inequality $f'_+(a) \leq 0$ will follow. When f is differentiable at the point a, one can also take the limit as $h \to 0$ along negative values and, in this case, $f'(a) = f'_-(a) = f'_+(a) \leq 0$.

Motivated by this simple calculus argument, consider now a continuously differentiable function f of another function $y(x)$, where y is defined on the interval $a < x < b$. Suppose that the integral

$$\int_a^b f(x, y(x)) \, dx \tag{9.1}$$

attains a maximum for some specific function $y_0(x)$ among all continuously differentiable and nonnegative $y(x)$ having the same values $y(a)$ and $y(b)$. We wish to find a condition on f that must hold at the maximum, similar to the derivative conditions that apply in the simpler scalar case. Just as we considered $f(a + h)$ for nonnegative scalar h earlier, we now write $f(x, y_0(x) + \varepsilon h(x))$ for $\varepsilon \geq 0$. We say that a continuously differentiable function h is an *admissible variation* if $h(a) = h(b) = 0$ and if the function $y_0 + \varepsilon h$ is nonnegative for all x in the interval whenever ε is small enough. In particular, we require that h be nonnegative for all x where $y_0(x)$ is zero (Figure 9.2).

The following Lemma will be useful in the subsequent section.

Lemma 9.1   *Let y_0 provide a maximum of the integral (9.1) among all continuously differentiable nonnegative functions y having the same values at a and b. Then*

$$
\begin{aligned}
f_y &= 0 \quad \text{for all x where $y_0(x) > 0$} \\
f_y &\leq 0 \quad \text{for all x where $y_0(x) = 0$}
\end{aligned}
\tag{9.2}
$$

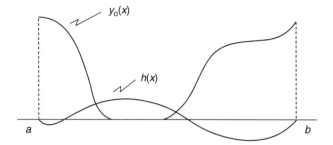

Figure 9.2. An admissible variation $h(x)$.

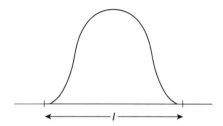

Figure 9.3. A smooth function $h(x)$ that is zero outside the interval I.

The subscript y denotes the derivative of f with respect to the argument y.

Proof: Choose an admissible variation h and write

$$F(\varepsilon) = \int_a^b f(x, y_0(x) + \varepsilon h(x))\, dx$$

F is real-valued and differentiable in ε, and it achieves a maximum at $\varepsilon = 0$. Taking the derivative of F with respect to ε inside the integral sign, our earlier discussion tells us that $F'(0) \le 0$. But

$$F'(\varepsilon) = \int_a^b f_y(x, y_0(x) + \varepsilon h(x)) h(x)\, dx$$

and so

$$\int_a^b f_y(x, y_0(x)) h(x)\, dx \le 0 \tag{9.3}$$

Let $y_0(x)$ be positive at some x in the interval. Then, because y_0 is a smooth function, it continues to be positive in some open interval I about x. Suppose that f_y is positive (negative) at x and choose h to be positive (negative) on I and zero elsewhere (Figure 9.3). If I is small enough, h is an admissible variation. In either case, (9.3) becomes an integral over I that is positive. This contradiction shows that f_y must be zero at x. On the other hand, if y_0 is zero at x, then choosing h to be positive on I forces f_y to be nonpositive there if we are not to contradict the inequality (9.3). This completes the proof.   □

9.3. Hide and Seek

A submarine is hiding somewhere within a region R of the ocean while its enemy engages in a search effort by patrolling the surface of R at random, either by plane or ship. For ease of presentation, we take R to be an interval of the real x axis.

The erratic movements of the searcher are designed to surprise the sub commander, who would otherwise be able to predict the whereabouts of the opposing forces, enabling him to position his vessel so as to minimize detection. The probability of detection, either by visual sighting or by a combination of radar and sonar, depends on the proximity of target and searcher. The amount of effort exerted in searching any given location, in terms of hours spent or miles covered per unit force, has a probability density function $g(x)$ that describes the frequency with which a point x is covered by the random patrol. Given that the target is actually at x, the conditional probability of detection is then $1 - e^{-g(x)}$. This result can be motivated by considering a discrete search consisting of n independent glimpses, each of which has a probability g of detecting the sub. Then the probability that the target escapes detection is, by the assumed independence, $(1 - g)^n$ and so the probability of being observed is one minus this quantity. When n is large we know from calculus that this expression can be approximated by an exponential: $1 - (1 - g)^n \approx 1 - e^{-ng}$. Because we are dealing here with a sequence of Bernoulli trials (success or failure of detection) the quantity ng is the average number of successes in n trials. In the case of continuous tracking, in which the glimpses are no longer discrete, ng is replaced by the mean number of sightings per unit time at position x, a quantity that is assumed to be proportional to the effort $g(x)$. Because visibility varies from place to place due to changes in daylight, depth, fog, obstacles, or some other interference, the function g is multiplied by a factor $a(x)$ between zero and one to indicate the efficiency of search at position x. The conditional probability of detection then becomes $1 - e^{a(x)g(x)}$ and therefore the unconditional probability of finding the hidden sub in the region R is

$$P = \int_R p(x)(1 - e^{-a(x)g(x)}) \, dx \qquad (9.4)$$

where $p(x)$ is the probability density function of the target being at x. The choice of p is up to the submarine commander, who picks this distribution so as to minimize the quantity P. The integrand in (9.4) expresses a "law of diminishing returns" in which a doubling of effort results in something less than a doubling of P, even when $a(x)$ is identically one.

The enemy searchers, on the other hand, want to maximize P provided that the cost of doing so is not exorbitant. Let $b(x)$ reflect the cost per unit effort of searching at x in terms of the difficulty or risk incurred. Then the total cost over R is

$$C = \int_R b(x)g(x)\,dx \tag{9.5}$$

and the searcher wants to maximize (9.4) while minimizing some multiple r of (9.5). A large r signifies a relatively high concern for cost, whereas a low value of r means that cost is relatively insignificant in terms of being able to find the target. Finding the minimum of C is the same as finding the maximum of $-C$, as a little reflection will reveal, and so the searcher's goal is to maximize a weighted sum $P - rC$.

The searcher must choose g while the opponent has the option of selecting p, and both do so optimally. What we wish to show next is how the two opposing players decide on their strategies.

We begin by maximizing $P - rC$ by choosing a smooth function g. Clearly g must be nonnegative and so Lemma 9.1 is applicable in which the function f is now

$$p(x)(1 - e^{-a(x)g(g)}) - rb(x)g(x)$$

and g plays the role of y. Relations (9.2) supply the rule

$$
\begin{aligned}
a(x)p(x)e^{-a(x)g(x)} - rb(x) = 0 \quad &\text{when } g(x) > 0 \\
a(x)p(x) - rb(x) < 0 \quad\quad\quad\quad &\text{when } g(x) = 0
\end{aligned}
\tag{9.6}
$$

If we denote $rb(x)/a(x)$ simply as $c(x)$ then, after a little bit of algebraic juggling, rule (9.6) becomes

$$
\begin{aligned}
g(x) &= 0 \quad\quad\quad\quad\quad\quad &&\text{if } p(x) \le c(x) \\
g(x) &= \frac{\ln p(x) - \ln c(x)}{a(x)} \quad &&\text{if } p(x) > c(x)
\end{aligned}
\tag{9.7}
$$

where ln is the natural logarithm. This gives the optimal choice of g in terms of a choice of p.

The sub commander wishes to minimize P or, equivalently, to maximize $1 - P$. Now

$$1 - P = \int_R p(x)e^{-a(x)g(x)}dx \tag{9.8}$$

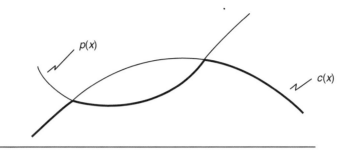

Figure 9.4. The minimum of the functions $p(x)$ and $c(x)$ shown as a dark curve.

since the integral of $p(x)$ over R is one (recall that p is a probability density). Using the rule (9.7) it is now easy to verify that

$$1 - P \leq \int_R c(x)\,dx \qquad (9.9)$$

and, indeed, if h is a function defined for each x by

$$h(x) = \min(p(x), c(x)),$$

then

$$1 - P = \int_R h(x)\,dx \qquad (9.10)$$

(Exercise 9.7.1 and Figure 9.4). It follows that $1 - P$ can never have a value that exceeds the integral over R of either $c(x)$ or $p(x)$, whichever is smaller.

The optimum strategy for the submarine is therefore now clear. It must choose $p(x)$ so that the area under the curve $h(x)$ is as large as possible. If the integral of the function $c(x)$ over R is less than unity then it suffices to choose $p(x)$ everywhere greater than $c(x)$, thereby giving $1 - P$ its largest value, namely, the integral of $c(x)$. On the other hand, if this integral exceeds one, then the optimum is obtained by letting $p(x)$ be everywhere less than $c(x)$, since this gives $1 - P$ a maximum value of one. It follows that no matter how the searcher picks g, the submarine can guarantee a minimum level of detection by properly choosing p.

The functions $a(x)$ and $b(x)$ are determined by external factors and are not at the discretion of either side. Therefore the value of $c(x)$ is determined solely by the constant r and the decision as to whether p is less than or greater than c in (9.7) depends on how small or large r is. Recall that a large r means that the probability of detection is less important than its cost and, conversely, when r is small, detection is paramount over cost.

We can now summarize the optimal tactics of each side in terms of r. When r is small:

$$g(x) = \frac{\ln(p(x)/c(x))}{a(x)}$$

$p(x)$ = any curve lying below the curve $c(x)$

and when r is large:

$$g(x) = 0$$
$$p(x) = \text{any curve lying above the curve } c(x)$$

Just how small is "small"? The answer is revealed in Exercise 9.6.2.

It is apparent from this discussion that if the cost of search is large enough in certain regions of R, then no effort should be expended there, and that when a search is undertaken, it is never proportional to the probability of finding the target. Also note that the optimal tactics for each side of the conflict are chosen on the assumption that the other side can pick its strategy with complete freedom. The pessimistic antagonists believe that the opponent will always choose g or p so as to thwart the other, and this leads both sides to pick their strategies so as to maximize what they are convinced is their minimum gain.

9.4. A Restricted Access Fishery

We now return to the fishery model of Section 7.2 and ask some new questions. The fishing grounds provide catches of several species that follow a growth law that we take to be logistic, in place of the form adopted earlier. If, as usual, x denotes the fish density and E the harvesting effort, then x satisfies the differential equation

$$x' = f(x) - \nu Ex = rx\left(1 - \frac{x}{K}\right) - \nu Ex \tag{9.11}$$

and the net revenue from the harvest is

$$E(\nu px - c) \tag{9.12}$$

All terms have the same meaning as before (we urge the reader to get re-acquainted with Section 7.2 before continuing).

Equation (9.11) has an equilibrium, for any fixed value of E, obtained by taking the intersection of the curve $f(x) = rx(1 - x/K)$ with the straight line νEx (Figure 9.5). It is an attractor. The total yield is given by the quantity νEx and so the maximum yield occurs at $x = K/2$, where $f(x)$ is greatest. However, if the equilibrium is at $x^* = c/\nu p$, and if this is less than $K/2$, as shown in Figure 9.5, the yield is also less.

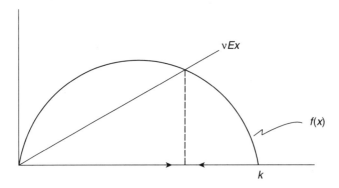

Figure 9.5. The logistic growth law $f(x)$ intersected by the line vEx. The point of intersection corresponds to an attractor and the yield from the harvest is the value of $f(x)$ at the attractor.

In an open-access fishery, each participant hastens to get what they can from the fishery by exploiting it to the utmost, because, as we explained earlier, to do otherwise would mean relinquishing his or her haul to the other fishermen. Investments in fishing gear and boats are largely irreversible, and so there is every incentive to continue fishing until the net revenue has been driven to zero. Denoting this revenue by $R(E)$ we see from (9.12) that $R(E) = 0$ precisely at $x = x^\star$.

Now consider the opposite situation in which the fishery has a single owner, a firm or some public entity. In contrast to the remorseless effort exerted in the common-access case, a sensible harvesting policy in the absence of competition would be to maximize the revenue R. Instead of trying to bring R down to zero, it might be wiser to restrain the fishing effort so that a positive net income can accrue over some indefinite time horizon. The goal of an individual owner is then to maximize the integral $J(E)$ defined by

$$J(E) = \int_0^\infty e^{-\delta t} R(E(t)) \, dt = \int_0^\infty e^{-\delta t} E(vpx - c) \, dt \qquad (9.13)$$

by choosing E properly as some function of time t. The factor $e^{-\delta t}$ is introduced to indicate that a future gain is worth less than a current profit by an exponential amount. δ is called a *discount factor*.

The maximum of (9.13) admits an easy solution. Assume, first, that the fishing effort E is constrained to have a maximum value E_m, corresponding to a limit on fleet size or total gear that can be deployed per unit time: $0 \le E \le E_m$. From (9.11) we see that $E = (f(x) - x')/vx$ and so (9.13) can be rewritten as

$$\int_0^\infty e^{-\delta t} \left(p - \frac{c}{vx} \right) f(x) \, dt - \int_0^\infty e^{-\delta t} \left(p - \frac{c}{vx} \right) x' \, dt \qquad (9.14)$$

Now define $Z(x)$ by the integral

$$Z(x) = \int_{x^\star}^{x} \left(p - \frac{c}{v} \right) du$$

Using an integration by parts, the second integral in (9.14) can be expressed as

$$\int_0^\infty e^{-\delta t} Z(x(t)) \, dt + Z(x(0))$$

where $x(0)$ is the initial value of x, which we assume to be greater than $x^\star$. As a matter of fact $x(t)$ will take on values only between K, its maximum population density, and $x^\star$, where the economic incentive to continue fishing is zero.

The expression for $J(E)$ is now

$$J(E) = \int_0^\infty e^{-\delta t} \left[\left(p - \frac{c}{vx} \right) f(x) - \delta Z(x) \right] dt \qquad (9.15)$$

The function $F(x) = (p - c/vx)f(x) - \delta Z(x)$ that appears in the integrand has a unique positive maximum $\hat{x}$ somewhere in the interval between $x^\star$ and K. To see this first note that for any given effort E one has $vEx = f(x)$ at an equilibrium x of (9.11). Therefore, letting $q(x) = f(x)(p - c/vx)$, it follows that $q(x) = E(vpx - c)$, which is the net revenue associated with the equilibrium value of x. As x increases from $x^\star$ the revenue increases to a maximum at some x_0 and then decreases. It follows that $q'(x)$ is monotone downward, crossing the x axis at x_0. To find the maximum of $F(x)$ we set F' to zero to obtain the equation $q'(x) = \delta(p - c/vx)$. The curves q' and $\delta(p - c/vx)$ intersect at some $\hat{x}$ that lies between $x^\star$ and x_0 (Figure 9.6).

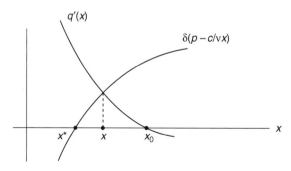

Figure 9.6. Intersection of the curves q' and $\delta(p - c/vx)$.

Note that as δ goes to infinity, $\hat{x}$ tends to $x^{\star}$. This is what one would expect in view of the fact that $\delta = \infty$ corresponds to completely discounting the future in favor of revenues that are earned immediately. This is the prevalent attitude in a common access fishery in which, as we have seen, revenue is driven to zero and the biomass level is reduced to $x^{\star}$.

Because $J(E)$ is to be maximized it is apparent that $x(t)$ should be chosen so that it reaches $\hat{x}$ from its starting point $x(0)$ as quickly as possible, and then remains there for all subsequent time. This is because e^{-t} is a decreasing function of t and so it is incumbent to reach the maximum value of $F(x)$ as soon as possible. If x is initially greater (lesser) than $\hat{x}$, then evidently x' should be chosen to have the largest possible negative (positive) value. It follows that the optimal harvesting policy for a single owner is

$$E(t) = \begin{cases} E_m & \text{if } x(t) > \hat{x} \\ 0 & \text{if } x(t) < \hat{x} \\ f(x)/vx & \text{if } x(t) = \hat{x} \end{cases} \tag{9.16}$$

Now suppose that the fishery is limited to a finite number of individuals or firms by the imposition of entry fees or licenses, either of which serves to restrict entry. For simplicity, we consider the case of only two owners who vie with each other to exploit the fishery. Each differs in the values of E, p, c, and v; we indicate this by the subscript $i = 1, 2$ to distinguish between them. The basic growth Equation (6.11) now becomes

$$x' = f(x) - v_1 E_1 x - v_2 E_2 x \tag{9.17}$$

because both owners ply the same waters, and the fish biomass is reduced by their combined efforts. The long-term revenue accrued by each owner is consequently a function of both harvesting rates and by analogy with (9.13) we write

$$J_i(E_1, E_2) = \int_0^{\infty} e^{-\delta t} (v_i p_i x - c_i) E_i \, dt \tag{9.18}$$

If each owner works separately, without competition from the other, then each would attain an optimal harvesting level $\hat{x}_i$, $i = 1, 2$, as explained earlier. But in reality they must they seek a competitive equilibrium that compromises the individual goals of maximizing (9.18). A pair of $E_i^{\#}$ will be deemed optimal if

$$J_1(E_1^{\#}, E_2^{\#}) \geq J_1(E_1, E_2^{\#})$$
$$J_2(E_1^{\#}, E_2^{\#}) \geq J_2(E_1^{\#}, E_2) \tag{9.19}$$

for all admissible E_i that satisfy $0 \leq E_i \leq \bar{E}_i$. This means that if owner 1, for example, unilaterally changes his strategy while owner 2 sticks to his, then J_1 either remains the same or, at worst, decreases.

Let $x_i^* = c_i/vp_i$. Then owner 1 is more efficient than owner 2 if $x_1^* < x_2^*$. This means that owner 1 operates at lower costs, or fetches higher prices for his catch, or is technologically better equipped. In this case, the competitive equilibrium is given by the rules

$$E_1^\#(t) = \begin{cases} \bar{E}_i & \text{if } x(t) > \eta \\ 0 & \text{if } x(t) < \eta \\ f(x)/v_1 x & \text{if } x(t) = \eta \end{cases} \qquad (9.20)$$

where $\eta = \min(\hat{x}_1, x_2^*)$, and

$$E_2^\#(t) = \begin{cases} \bar{E}_i & \text{if } x(t) \geq x_2^* \\ 0 & \text{if } x(t) < x_2^* \end{cases} \qquad (9.21)$$

Let us establish the truth of (9.20). The other rule (9.21) is verified in a similar fashion (Exercise 9.6.4). To begin with, define $f_1(x)$ by

$$f_1(x) = \begin{cases} f(x) - v_2 E_2 x & \text{if } x \geq x_2^* \\ f(x) & \text{if } x < x_2^* \end{cases}$$

If $\bar{E}_2$ is large enough, f_1 is negative for $x \geq x_2^*$. Now assume that rule (9.21) holds. Then (9.17) becomes

$$x' = f_1(x) - v_2 E_2 x$$

By analogy with the way $F(x)$ was defined earlier, let

$$F_1(x) = \left(p_1 - \frac{c_1}{v_1 x} \right) f_1(x) - Z_1(x)$$

in which Z_1 is the same integral as Z except that its lower limit is x_1^*. It is always true that $\hat{x}_1 > x_1^*$. Moreover, because $x_1^* < x_2^*$, Figure 9.7 shows that F_1 is maximized by choosing x equal to the smallest of $\hat{x}_1$ and x_2^* because $F_1(x)$ is immediately negative as soon as x exceeds x_2^*. The form of F_1 in this figure is arrived at from the same considerations that applied to F earlier. We conclude that $J_1(E_1, E_2^\#)$ must be maximized by rule (9.20), for the reason that $J(E)$ for a single owner was maximized previously using (9.16).

The interpretation of this result is that if owner 1 is more efficient then he drives owner 2 out of business since x is quickly brought down to a value at which the

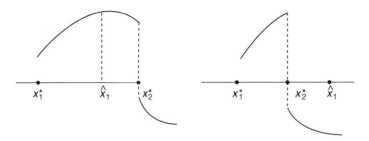

Figure 9.7. Maxima of the function $F_1(x)$ in two cases.

net revenue to 2 is zero. Since $\eta > x_1^*$, the revenue garnered by 1 is positive. When $\hat{x}_1 > x_2^*$ owner 1 eliminates his competitor by fishing more intensely than if this inequality is reversed. In effect the first owner must work hard to keep his competitor from returning. Finally, in the case in which $x_1^* = x_2^*$ then the revenues of both parties are evidently dissipated to zero and the situation is no better than in an open access fishery.

In summary, we see that if the goal of limited access is to safeguard against overfishing, it is not a very successful ploy because the intense competition between participants ensures that the fish stocks will be reduced as far as necessary to achieve the complete elimination of those who are less efficient. This may be recognized as another form of the principle of competitive exclusion that we encountered in Chapter 6. Evidently some other form of control, such as catch quotas or a tax on the harvest, is required for effective conservation.

9.5. A Comment about Strategy

It is interesting to compare the optimal strategies adopted by the various participants in the two previous sections. We have no intention of delving into the formalism of game theory, where these matters are discussed in more detail, but only wish to comment, in a slightly more abstract way, on the different approaches that were used.

There are two competitors C_i, $i = 1, 2$, who aim to achieve certain goals but are constrained by the presence of the other. The naval officers of the opposing forces, for example, have the goals of evading detection to launch torpedoes, on one side, and of locating the submarine and preventing it from completing its mission, on the other.

Let $G_1(u, v)$ and $G_2(u, v)$ represent the loss to each side, where u and v are strategies that are available to C_1 and C_2, respectively. In the submarine problem, G_i are measures of detection and evasion, as you recall, and u and v describe how searcher and target place themselves in the ocean.

If C_1 is pessimistic, as in the submarine problem, he believes that C_2 will try to thwart him by choosing a strategy v that will maximize the loss G_1. Therefore C_1

adopts a conservative attitude and picks u to minimize his maximum loss. The other competitor, equally suspicious, behaves similarly. Letting u_0 and v_0 be a pair of optimal strategies for each side, this means that

$$G_1(u_0, v_0) \geq G_1(u_0, v)$$
$$G_2(u_0, v_0) \geq G_2(u, v_0)$$

(9.22)

To see how (9.22) comes about, note that $G_1(u, v_0) = \max G_1(u, v) \geq G_1(u, v)$ for all u and v (the maximum is over all v) and, therefore, $G_1(u_0, v_0) = \text{minmax}$ $G_1(u, v) \geq \min G_1(u, v) = G_1(u_0, v)$, with the minimum taken over all u. A similar set of relations is applicable to G_2.

The first inequality in (9.22) expresses the idea that no matter what v is picked by C_2, the loss to C_1 can never be greater than $G_1(u_0, v_0)$ as long as C_1 chooses u_0. The second inequality has a similar interpretation. This is known as a *minmax strategy* and it is the one adopted by the submarine commanders.

What is being ignored here, of course, is the question as to whether optimal strategies exist to begin with, a mathematical issue that belongs to the domain of game theory.

A different competitive approach is the one adopted by the two owners of a fishery, in which both sides select strategies u_0 and v_0 so that any unilateral change on the part of C_1, say, to some other choice of u can only increase the loss G_1, as long as C_2 sticks to using v_0. Similar considerations apply to C_2. This can be expressed as a pair of inequalities:

$$G_1(u_0, v_0) \leq G_1(u, v_0)$$
$$G_2(u_0, v_0) \leq G_2(u_0, v)$$

(9.23)

This is referred to as a *Nash competitive strategy*.

A comparison of these approaches shows that minmax and competitive strategies need not equal each other because one is based on pessimism and the other on greed. There is one situation in which the two approaches are equal, however. Suppose that $G_1(u, v) = -G_2(u, v)$ so that a gain to one side is a loss to the other, as would be the case, for example, if the parameter r in the submarine problem is set to zero. Then (9.22) and (9.23) reduce to the same strategy pair (Exercise 9.6.7).

Although it plays no role in the models discussed in this chapter, there is yet another form of equilibrium in which the two players decide to cooperate by helping one another to a disadvantage to themselves. In this way, it is conceivable that the loss to each could be greater than if they opted for a noncooperative Nash strategy. A cooperative strategy can be defined as a pair u_0, v_0 for which there is no other u, v that satisfy the pair of relations

$$G_1(u_0, v_0) \geq G_1(u, v)$$
$$G_2(u_0, v_0) \geq G_2(u, v)$$

with at least one of these inequalities being strict. What this means is that any move away from an optimal pair of strategies can make only one of the players worse off.

9.6. Exercises

9.6.1. Establish the identity (9.10).

9.6.2. In the statement of an optimal strategy for the two naval opponents in Section 9.3 we distinguished between "small" and "large" r. Show that the difference between small and large occurs at the critical value

$$r = \frac{1}{\int_R (b(x)/a(x))\, dx}$$

9.6.3. A submarine wishes to pass through a channel of length L undetected by an airplane that patrols back and forth across the channel perpendicular to the path of the boat. This is called a barrier patrol. The sub cannot be detected when submerged, but it can remain under the water for only a maximum distance of $d < L$. The channel has a varying width and if the plane patrols the widest part, it cannot do so as efficiently as when it patrols the narrower parts. Nevertheless, the plane varies the position of the patrol at random every day to avoid giving the submarine commander the knowledge he needs to dive under the barrier and elude detection. The probability density of the position where the patrol takes place is given by $g(x)$. The integral of g over the length of the channel is one.

The situation is illustrated in Figure 9.8. If the patrol is at position x and the boat attempts to go by on the surface, then the conditional probability of detection is $p(x)$, where p is smaller where the channel is wide and larger where the channel is narrow.

Let $h(x)$ denote the probability that the submarine is submerged at position x. A little thought will convince us that the integral of h over the length of the channel

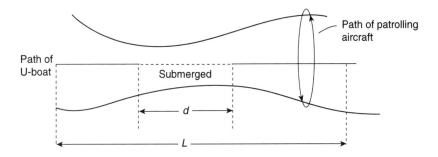

Figure 9.8. A barrier patrol in a channel of length L and varying width.

is d. We allow the sub to submerge for a distance less than d and then resubmerge elsewhere, as he chooses, provided that the total distance sums to d. The plane wishes to choose the function g to maximize the probability of detection, no matter what the sub does (a maxmin strategy), whereas the sub wants to maximize the probability of not being detected, regardless of what the plane does, by picking the function h. This too is a maxmin strategy. Formulate this problem mathematically and make an attempt at a solution.

9.6.4. Establish rule (9.21) by assuming that (9.20) holds and then showing that $J_2(E_1^\#, E_2)$ is maximized by (9.21).

9.6.5. A version of the model in Section 9.3 can be formulated as an optimization problem that is reminiscent of the type considered in Chapters 2 and 3. Suppose that the region R of the ocean is partitioned into N subsets, and that the probability the submarine is within the ith subset is a fixed number $p_i, i = 1, 2, \ldots, N$, where the p_i sum to unity. An amount of search effort g_i is allocated to region i, in terms of fuel, aircraft, time, or some other limited resource. The sum of the g_i is some stipulated constant that represents the total amount of available resources that can be expended in the search. The conditional probability of a sighting in region i, given that the target is actually within this area, is $1 - e^{-a_i g_i}$, where a_i is the visibility factor in area i (recall the discussion in Section 9.3). The goal is to maximize the probability of detection within R by choosing the deployments g_i optimally. Formulate this problem mathematically.

9.6.6. Another application of search theory arises in police work. Suppose that a police vehicle is assigned a sector or "beat" in which there are c street miles that are accessible to patrol and that the average speed is v miles per hour. In time t, the vehicle moves a distance vt at random. This means that the movement is generally zigzag through the streets to reduce any element of predictability on the part of a criminal. For small enough t, it is unlikely that the patrol car will overlap the path it just covered, however. The small distance vt is randomly located among the c miles of street network and a crime incident is also assumed to be located at random within the same c miles (thefts, muggings, and the like). Therefore the probability that in time t the path of the random patrol overlaps the location of a crime in progress is simply vt/c. This is the conditional probability of detecting a crime, given that there is a crime in progress during time t. As t increases, so does the likelihood that the patrol car will retrace part of its own path, and so this simple ratio is no longer an accurate description of the detection probability. However, one can reason as follows. Let $P(t)$ be the probability of no detection during t. Increase t by a small increment Δt. Nonoverlapping time intervals are assumed to give independent events and so it follows that $P(t + \Delta t) = P(t)P(\Delta t)$. However, $P(\Delta t) = 1 - (vt/c)$ and by letting t go to zero, one shows that P satisfies the differential equation $P' = 1 - (v/c)P$. Verify this and establish that the conditional probability of detecting a crime in progress during any time t is $1 - e^{-vt/c}$. This is similar to the law of diminishing returns that we encountered in the search for submarines in Section 9.3.

9.6.7. Use the minmax inequalities (9.22) to imply the Nash strategies (9.23) whenever $G_1(u, v) + G_2(u, v) = 0$, the so-called zero-sum situation.

9.7. Further Readings

Although the models in this chapter are a bit contrived and less convincing than the ones that were considered previously, it is nevertheless true that they reflect real issues. The hide-and-seek game, for example, was played out many times during the Second World War and in the annals of the military operations of that time there are tales reminiscent of the one we looked at (see, for example, the book [50]) and the restricted access fishery issue has been confronted by several nations. Differential games of the type discussed in Section 9.4 are treated in the book by Issacs [30].

Our discussion of the limited entry fishery is based on Clark's paper [22], whereas the search problem is adapted from [35]. The topic of search theory has wider applicability than just hunting for submarines and has been applied to police patrols and a variety of navy and coast guard search missions (see [20] and Exercise 9.6.6).

The barrier patrol problem is taken from [39].

Afterthoughts on Modeling

The goal of modeling is to gain insight into some problem that occurs in the natural world of real events. In a few cases, there is sufficient data and enough understanding of the actual processes that it makes sense to test the model against numbers obtained from the field and to make predictions that can be verified in a relatively unambiguous manner, as in the crab model of Chapter 1. In other cases, this is not possible either because a trustworthy database is too meager, or doesn't exist or, as is the case in the restricted access fishery model of Chapter 9, because an understanding of the underlying dynamics is inadequate. In the latter case, a model can still be useful, however, in providing a plausible metaphor for the observable behavior and even to suggest what the actual poorly known dynamics might be. The model can then be tested in some gross manner and provide direction for future investigations. Both kinds of models appear in this book, as you have seen.

Certain themes recur in different settings in various chapters showing that a common problem can be viewed from varying angles, depending on the questions asked. Even if they are handled by different methods there is at least an undercurrent of commonality. An example is the behavior of a blood-clotting cascade in Chapter 6 and the algae bloom problem of Chapter 7. Each describes an excitable system that is activated when certain parameters exceed a certain threshold. In one case, the equations describe biochemical reactions and, in the other, reactions among cellular organisms. There is a common thread linking the two phenomena. Strengthening this bond is the measles epidemic model of Chapter 6, which also displays threshold behavior, and so the fact that each similar differential equation model was put to work in each case should not be too much of a surprise.

Multiple and conflicting objectives in societal problems provide another example of commonality. Issues of this type came up in Chapters 2, 3, 4, and 9, and in each instance, optimization schemes were employed to penetrate the problems. In retrospect, this is again not unexpected.

Although most problems in this book are temporal in nature, sometimes spatial considerations cannot be avoided, as in Chapter 8. Similarly, most models are deterministic in nature although more difficult stochastic tools occasionally come into play, as in Chapter 3. Indeed, temporal, spatial, and stochastic considerations all appear in our discussion of models for the spread of a disease, from the detailed treatment of Chapter 6 through the Exercises 3.6.9, 3.6.10, and 8.6.5.

Depending on the questions you ask, one formulation of a problem may be more appropriate than another, and if one is to make any headway at all it is wise to ask modest questions at first to avoid being overwhelmed. In this book, simplifying assumptions are made in each chapter in the hope that the phenomenon we are looking at is robust enough to exhibit behavior that is qualitatively similar to the real thing even though it has been stripped of much of its complexity. We need to be aware of our assumptions and take that into account when we attempt to interpret what the models tell us. I try to do that in this book, but inevitably there are hidden factors that one is only dimly aware of, and this can lead to misleading conclusions. That, unfortunately, is one of the pitfalls of modeling.

In this book, the mathematics are driven by the problems, not the other way around. That is why we begin each chapter with a description of the problem setting and then suggest a likely approach. Good models begin this way, and to do otherwise is to engage in an exercise of trying to tailor a problem to fit the mathematics. This is another modeling pitfall that should be avoided.

Appendix. Conditional Probability

A few facts concerning conditional probability needed in Chapter 3 are reviewed here. For further details, consult the first three chapters of the book by Ross [46].

Let A and B denote events in a sample space S. The *conditional probability* of A given B, written $\text{prob}(A \mid B)$ is defined by

$$\text{prob}(A \mid B)\text{prob}(B) = \text{prob}(AB) \tag{A.1}$$

where AB is the joint event of A and B. When $\text{prob}(A \mid B) = \text{prob}(A)$, we say that A and B are independent.

As an example, suppose that X and Y are discrete random variables taking on nonnegative integer values. If A is the event "$X = k$" and B is "$Y = i$," then (A.1) becomes

$$\text{prob}(X = k \mid Y = i)\,\text{prob}(Y = i) = \text{prob}(X = k, Y = i)$$

Let B_i be a collection of disjoint events indexed by i, whose union is S. Then

$$\text{prob}(A) = \sum \text{prob}(AB_i) \tag{A.2}$$

(From now on, all sums are taken over the indicated index from 0 to ∞.)

In terms of X and Y, (A.2) means that

$$\text{prob}(X = k) = \sum \text{prob}(X = k, Y = i)$$

Because of (A.1), we can now write (A.2) as

$$\text{prob}(A) = \sum \text{prob}(AB_i)\,\text{prob}(B_i) \qquad (A.3)$$

and so, for the variables X and Y, we obtain

$$\text{prob}(X = k) = \sum \text{prob}(X = k \mid Y = i)\,\text{prob}(Y = i)$$

The *expected value* (also called the mean value) of X is defined by

$$E(X) = \sum k\,\text{prob}(X = k)$$

and if $h(X)$ is some function of X, then the expected value of the random variable $h(X)$ is given by

$$E(h(X)) = \sum h(k)\,\text{prob}(X = k) \qquad (A.4)$$

For example, if $h(X) = X^2$, then

$$E(X^2) = \sum k^2\,\text{prob}(X = k)$$

The *conditional expectation* (conditional mean) of X, given that $Y = i$, is defined by

$$E(X \mid Y = I) = \sum k\,\text{prob}(X = k \mid Y = i) \qquad (A.5)$$

Relation (A.5) enables us to define $E(X \mid Y)$ as a function of Y; call it $h(Y)$, whose value when $Y = i$ is given by (A.5). From (A.3) and (A.4), we therefore obtain the unconditional expectation of X as

$$E(E(X \mid Y) = \sum h(i)\,\text{prob}(Y = i) = \sum E(X \mid Y = i)\,\text{prob}(Y = i)$$

$$= \sum \sum k\,\text{prob}(X = k \mid Y = i)\,\text{prob}(Y = i)$$

$$= \sum k \sum \text{prob}(X = k, Y = i) = \sum k\,\text{prob}(X = k)$$

It follows that

$$E(X) = E(E(X \mid Y)) \qquad (A.6)$$

When X and Y are continuous random variables taking on real nonnegative values, the sums are replaced by integrals over a continuum of events indexed by the nonnegative real numbers. Moreover, the discrete probabilities $\text{prob}(X = k)$ are now represented by a continuous density function $f(s)$. Consider, for example,

X and Y to be exponentially distributed random variables (see Section 3.2). The event "$X < Y$" means that the values assumed by X are less than the values taken on by Y. Then (A.2) and (A.3) become

$$\text{prob}\,(X < Y) = \int \text{prob}\,(X < Y \mid Y = s)\,ds$$

and

$$\text{prob}\,(X < Y) = \int \text{prob}\,(X < Y \mid Y = s)\mu e^{-\mu s}\,ds$$

(From now on, all integrals are taken from 0 to ∞.)

Relation (A.6) is now expressed as

$$E(X) = E(E(X \mid Y)) = \int E(X \mid Y = s)\mu e^{-\mu s}\,ds$$

References

1. Anderson, D. "Red Tides." *Scientific American*, August, 1994.
2. Anderson, R., and R. May. *Infectious Diseases of Humans*. Oxford Press, 1992.
3. Aumann, R., and M. Maschler. "Game Theoretic Analysis of a Bankruptcy Problem in the Talmud." *J. Economic Theory* 36, 1985, 195–213.
4. Bailey, N. *The Mathematical Theory of Infectious Diseases*. Charles Griffin, 1975.
5. Balinski, M., and P. Young. *Fair Representation*. Yale University Press, 1982.
6. Bauer, W., F. Crick, and J. White. "Supercoiled DNA." *Scientific American*, July, 1980.
7. Beltrami, E. "Unusual Algal Blooms as Excitable Systems: The Case of the Brown Tide." *Envir. Modeling Assessment* 1, 1996, 19–24.
8. Beltrami, E., and J. Jesty. "Mathematical Analysis of Activation Thresholds in Enzyme-Catalyzed Positive Feedbacks: Application to the Feedbacks of Blood Coagulation." *Proc. Nat. Acad. Sciences* 92, 1995, 8744–8748.
9. Beltrami, E., and O. Carroll. "Modeling the Role of Viral Disease in Recurrent Phytoplankton Blooms." *J. Math. Biology* 32, 1994, 857–863.
10. Beltrami, E. *Mathematics for Dynamic Modeling*. Academic Press, 1998.
11. Beltrami, E., and L. Bodin. "Networks and Vehicle Routing for Municipal Waste Collection." *Networks* 4, 1974, 65–94.
12. Beretta, E., and Y. Kuang. "Modeling and Analysis of a Marine Bacteriophage Infection." *Math. Biosciences* 149, 1998, 57–76.
13. Blumstein, A., and R. Larson. "Problems in Modeling and Measuring Recidivism." *J. Research in Crime and Delinquency*, 1971, 124–132.
14. Bodin, L. "Towards a General Model for Manpower Scheduling, Parts 1 and 2." *J. Urban Analysis* 1, 1973, 191–208 and 223–246.
15. Braun, M., C. Coleman, and D. Drew, eds. *Differential Equation Models*. Springer-Verlag, 1983.
16. Burke, P. "The Output of a Queueing System." *Operations Research* 4, 1958, 669–704.
17. Carter, G., J. Chaiken, and E. Ignall, eds. *Fire Department Deployment Analysis*. Elsevier North-Holland, 1979.
18. Chaiken, J. "The Number of Emergency Units Busy at Alarms Which Require Multiple Servers." Rand Report R-531-NYC/HUD, 1971.
19. Chase, I. "Vacancy Chains." *Annual Review of Sociology* 17, 1991, 133–154.
20. Chelst, K. "The Basics of Search Theory Applied to Police Patrols." In *Police Deployment*, R. Larson, ed. D.C. Heath, 1978, 161–182.

21. Cipra, B. "Mathematics Untwists the Double Helix." *Science* 247, 1990, 913–915.
22. Clark, C. "Restricted Access to Common Property Fishery Resources: A Game-Theoretic Analysis." In *Dynamic Optimization and Mathematical Economics*. Plenum Press, 1980, 117–132.
23. Clark, C. *Mathematical Bioeconomics*. John Wiley, 1976.
24. Courant, R., and F. John. *Introduction to Calculus and Analysis*, Vol. 2. John Wiley, 1974.
25. Crutchfield, J., J. Farmer, N. Packard, and R. Shaw. "Chaos." *Scientific American*, December, 1986.
26. D'Ancona, U. *The Struggle for Existence*. E. Brill, 1954.
27. Frauenthal, J. *Smallpox*. Birkhauser, 1981.
28. Gould, S. J. "Unenchanted Evening." In *Eight Littles Piggies*. Norton, 1993.
29. Hardin, G. "The Tragedy of the Commons." *Science* 162, 1968, 1243–1248.
30. Issacs, R. *Differential Games*. John Wiley, 1965.
31. Jeffrey, Alan. *Advanced Engineering Mathematics*. Academic Press, 2001.
32. Kemeny, J., and L. Snell. *Finite Markov Chains*. D. Van Nostrand, 1960.
33. Kemeny, J., and L. Snell. *Mathematical Models in the Social Sciences*. Ginn, 1961.
34. Kolesar, P., and E. Blum. "Square Root Laws for Fire Engine Response Distances." *Management Science* 19, 1973, 1368–1378.
35. Koopman, B. *Search and Screening*. Pergamon Press, 1980.
36. Lea, D. E., and C. Coulson. "The Distribution of the Numbers of Mutants in Bacterial Populations." *J. Genetics* 49, 1949, 264–285.
37. Maltz, M. *Recidivism*. Academic Press, 1984.
38. Mechling, J. "A Successful Innovation: Manpower Scheduling." *J. Urban Analysis 2*, 1974, 259–313.
39. Morse, P., and G. Kimball. *Methods of Operations Research*. John Wiley, 1951.
40. Murray, J. *Mathematical Biology*. Springer Verlag, 1989.
41. O'Neill, B. "A Problem of Rights Arbitration from the Talmud." *Mathematical Social Sci.* 2, 1982, 345–371.
42. Okubo, A. *Diffusion and Ecological Problems*. Springer-Verlag, 1980.
43. Okubo, A., P. Maini, M. Williamson, and J. Murray. "On the Spatial Spread of the Gray Squirrel in Britain." *Proc. Royal Soc. London* B 238, 1989, 113–125.
44. Olsen, L., and W. Schaeffer. "Chaos Versus Noisy Periodicity: Alternative Hypothesis for Childhood Epidemics." *Science* 249, 1990, 499–504.
45. Pool, R. "Is It Chaos or Is It Just Noise ?" *Science* 243, 1989, 25–28.
46. Ross, S. *Introduction to Probability Models*. Academic Press, 1990.
47. Slobodkin, L., and H. Kierstead. "The Size of Water Masses Containing Plankton Blooms." *J. Marine Research* 12, 1953, 141–157.
48. Tuchinsky, P. *Man in Competition with the Spruce Budworm*. Birkhauser, 1981.
49. Tucker, A. "Perfect Graphs and an Application to Optimizing Municipal Services." *SIAM Review* 15, 1973, 585–590.
50. Waddington, C. *OR in World War II: Operations Research Against the U-Boat*. Elek, 1973.
51. Weissburg, M., L. Rosemann, and I. Chase. "Chains of Opportunity: A Markov Model for the Acquisition of Reusable Resources." *Evolutionary Ecology* 5, 1991, 105–117.
52. White, H. *Chains of Opportunity: Systems Models of Mobility in Organizations*. Harvard Press, 1970.
53. White, J. "Self-linking and the Gauss Integral in Higher dimensions." *Amer. J. Math.* 91, 1969, 693–728.
54. White, J., and W. Bauer. "Calculation of the Twist and the Writhe for Representative Models of DNA." *J. Molecular Biology* 189, 1986, 329–341.

Solutions to Select Exercises

Chapter 1

1.5.2 From the previous exercise, $E(N_i) = (1 - f_i) - 1 = t_{ii}$. But $t_{ii} = 1 + h_{ii}t_{ii}$. Thus, $h_{ii} = (t_{ii} - 1)/t_{ii} = 1 - (1 - f_i) = f_i$.

1.5.4 $t_{3,2} = p/(1 - p) = h_{3,2}t_{2,2}$. But $t_{2,2} = 1/(1 - p)$ and therefore $h_{3,2} = p = .9$. Also, $t_{4,4} = (1 - 15p/16)/(1 - p) = 10(2.5/16) \cong 1.56$.

1.5.8 p_i, s_i, and h_i do not depend on past moves but only on the current state i. Moreover, they do not depend on the year in which the evaluations take place. If $q_4 = 1$ (Professor Emeritus), then the chain is absorbing. The transition matrix **P** is given below

$$
\begin{array}{c}
 \\
1 \\
2 \\
3 \\
4
\end{array}
\begin{array}{c}
\begin{array}{cccc}
1 & 2 & 3 & 4
\end{array} \\
\left(
\begin{array}{cccc}
q_1 + h_1 & p_1 & 0 & 0 \\
h_2 & q_2 & p_2 & 0 \\
h_3 & 0 & q_3 & p_3 \\
h_4 & 0 & 0 & q_4
\end{array}
\right)
\end{array}
$$

Chapter 2

2.5.5 $C(N) = \sum x_i$, summed from 1 to N, and so $\sum x_i \le C(N)$ when summed over any subset S of i values. The sum over i in S plus the sum over i in $N - S$ is $C(N)$ from which it follows that $\sum x_i \le C(N - S)$ or, put another way, $C(N - S) + \sum x_i \leq C(N)$, where the sum is again over S. On the other hand, if the sum of x_i over S is not less than $C(N) - C(N - S)$, then the sum of x_i over $N - S$ is no less than $C(N) - C(S)$, namely, $C(S) \geq C(N) - \sum x_i$ with the sum over i now in $N - S$. Therefore, $C(S)$ minus the sum of x_i over S is no less than $C(N)$ minus the sum of x_i over all i, namely, zero.

2.5.6 We wish to minimize the sum $x_1 + x_2 + x_3 + x_4$ subject to $Ax \geq n$, where

$$A = \begin{pmatrix} 1 & 0 & 0 & 1 \\ 1 & 0 & 0 & 0 \\ 0 & 1 & 0 & 0 \\ 0 & 0 & 1 & 0 \\ 0 & 0 & 1 & 1 \end{pmatrix} \qquad n = \begin{pmatrix} 6 \\ 4 \\ 5 \\ 6 \\ 8 \\ 10 \end{pmatrix}$$

The solution is $x_1 = 4$, $x_2 = 6$, $x_3 = 8$, $x_4 = 2$ and the sum of the x_i from 1 to 4 equals 20. Each person works 5 days out of 7. If the total workforce is N then $N - 20$ are allowed off each day. Therefore the sum of $(N - 20)$ over 7 days must equal $2N$, namely, $N = 28$.

2.5.7 Suppose that an estate E has two allocations $x_1 \leq x_2 \leq x_3$ and $y_1 \leq y_2 \leq y_3$ and that $x_i < y_i$ for some i. If $x_j \leq y_j$ for all $j \neq i$, then $E = x_1 + x_2 + x_3 < y_1 + y_2 + y_3 = E$, which is impossible. Therefore,

$$x_j > y_j \qquad \text{and} \qquad x_i \leq y_i, \quad \text{for some } j \neq i \qquad (*)$$

Let S be a subset of contestants containing only i and j. Then S gets an allocation of E_S equal to either $x_i + x_j$ or $y_i + y_j$. Suppose that $x_i + x_j \leq y_i + y_j$. Internal consistency means that under the allocation $x_i + x_j$ the ith contestant gets x_i and the jth gets x_j, whereas under an allocation of $y_i + y_j$ the ith person gets y_i and the jth receives y_j. Because of monotonicity, y_j must not be less than x_j, which contradicts $(*)$. A similar argument shows that $(*)$ is again contradicted whenever $y_i + y_j < x_i + x_j$. It follows that $x_i = y_i$ for all i.

Chapter 3

3.6.3 $F(r) = \text{prob}(D_1 \leq r) = 1 - e^{-\gamma \pi r^2}$ and so $E(D_1) = \int_0^\infty r F'(r) \, dr = -2\gamma d/d\gamma (\int_0^\infty e^{-\gamma \pi r^2} \, dr)$. Make the change of variable $x = (\gamma \pi)^{1/2} r$ to obtain $E(D_1) = -\gamma d/d\gamma (1/\sqrt{\gamma}) = 1/2\sqrt{\gamma}$.

3.6.6 Let $h(q) = E(D_1 | q)$, where $q = N - m$ and $q_o = E(q) = N - \lambda/\mu$. From the convexity of $h(q)$ it follows that $h(q) \geq h(q_o) + (q - q_o)h'(q_o)$. Taking the expected value of both sides of this inequality gives

$E(D_1) = E(E(D_1|q)) = E(h(q)) \geq h(E(q)) + h'(q_o)E(q - q_o)$. But $E(q - q_o) = E(q) - E(q) = 0$ and so $E(h(q)) \geq h(E(q))$.

3.6.7 Let L be the number of busy fire companies and M the number of un-covered response neighborhoods. We wish to choose x_j as zero or one so that the sum of the x_j from 1 to L is minimized, subject to having $\sum a_{ij}x_j \geq 1$, for $I = 1, 2, \ldots, M$, and where the sum extends from 1 to L.

3.6.10 $E(W_m) = E(E(W_m \mid W_{m-1})) = E(\theta W_{m-1}) = \theta E(W_{m-1})$. Therefore, $E(W_1) = \theta$, $E(W_2) = \theta^2, \ldots, E(W_m) = \theta^m$. Because $\theta < 1$, $\sum \theta^m = 1/(1 - \theta) = \mu/(\lambda - \mu)$, where the sum is over all m from 0 to infinity. But $E(W_m) = \sum k \, \text{prob}(W_m = k) \geq \sum \text{prob}(W_m = k) = \text{prob}(W_m \geq 1)$; the sums extend from 1 to infinity. Because $\theta^m \to 0$, it follows that $\text{prob}(W_m = 0) = 1 - \text{prob}(W_m \geq 1) \to 1$, as $m \to \infty$.

Chapter 4

4.5.3 The sum of the inner degrees must equal the sum of the outer degrees, over all nodes, since the inner and outer degrees are the same, node by node. It follows from Lemma 4.1 that $2e = \sum \delta(v_i) = \sum \delta_+ + \sum \delta_- = 2\sum \delta_+$, which is what we needed to show.

4.5.4 Let n_o be the number of odd degree nodes. If n_o is itself an odd number, then the sum of $\delta(v_i)$ over the odd degree nodes is odd, since it is an odd sum of odd numbers. This contradicts Lemma 4.2

4.5.7 We need to minimize $\sum x_j$ summed over j from 1 to M, subject to $\sum a_{ij}x_j \geq 1$, where $i = 1, 2, \ldots, N$, and the x_j are zero or one variables.

Chapter 5

5.4.3 The curves C_1 and C_2 are parametrized by $\mathbf{r}_1(s) = (\cos\theta, \sin\theta, 0)$ and $\mathbf{r}_2(\tau) = (0, 0, \tau)$, respectively, and $\mathbf{r} = \mathbf{r}_2 - \mathbf{r}_1 = (\cos\theta, \sin\theta, -\tau)$. Therefore $\mathbf{T}_1 = (-\sin\theta, \cos\theta, 0)$ and $\mathbf{T}_2 = (0, 0, 1)$ and $\mathbf{T}_2 \times \mathbf{T}_1 = (\cos\theta, \sin\theta, 0)$. Since $\mathbf{v} = \mathbf{r}/\gamma^{3/2}$ with $\gamma = (1 + \tau^2)^2$, it follows that $\mathbf{v} \cdot (\mathbf{T}_2 \times \mathbf{T}_1) = (1 + \tau^2)^{-3/2}$. Expression (5.7) becomes

$$\frac{1}{4\pi} \iint \frac{d\theta d\tau}{(1 + \tau^2)^{3/2}}$$

where θ ranges from 0 to 2π and τ from $-\infty$ to $+\infty$. Let $\tau = \sinh u$ and note that $d\tau/du = \cosh u$. Under this change of variable the integral transforms to

$$\int \frac{du}{\cosh^2 u}$$

where the integral ranges from 0 to ∞. However, $d(\tanh u)/du = 1/\cosh^2 u$ and so the integral is readily evaluated to have the value $+1$.

5.4.4 If curve C lies in the x, y plane, then so does its tangent and therefore $\mathbf{T}(s) = (x'(s), y'(s), 0)$. It follows that $d\mathbf{T}/ds = (x'(s), y'(s), 0)$ lies in the same plane. Since $\mathbf{B}$ is orthogonal to $\mathbf{T}$ and $\mathbf{N}$, it points in the z direction with constant length of 1. That is, $\mathbf{B} = (0, 0, 1)$ and so $d\mathbf{B}/ds = 0$ along C. Since $\mathbf{B}$, $\mathbf{T}$, and $\mathbf{N}$ are orthogonal, $0 = d(\mathbf{B} \cdot \mathbf{N})/ds = \mathbf{B}' \cdot \mathbf{N} + \mathbf{B} \cdot \mathbf{N}' = \mathbf{B} \cdot \mathbf{N}'$, from which it follows that $\mathbf{N}'$ is orthogonal to $\mathbf{B}$. $\mathbf{N}'$ is also orthogonal to $\mathbf{N}$ since it has constant unit length along C. Therefore, $\mathbf{N}'$ is some multiple β of $\mathbf{T}$. Now, $0 = d(\mathbf{N} \cdot \mathbf{T})/ds = \mathbf{N}' \cdot \mathbf{T} + \mathbf{N} \cdot \mathbf{T}' = \mathbf{N}' \cdot \mathbf{T} + k$ (since $\mathbf{T}' = k\mathbf{N}$); we see from this that $\mathbf{N}' \cdot \mathbf{T} = -k$. Using the fact that $\mathbf{N}' = \beta\mathbf{T}$ we get that $\beta = -k$ and therefore $\mathbf{N}' = -k\mathbf{T}$, as required.

5.4.5 $\mathbf{V}'(s) = -\alpha \sin(\alpha s)\mathbf{N} + \cos(\alpha s)\mathbf{N}' + \alpha \cos(\alpha s)\mathbf{B} + \sin(\alpha s)\mathbf{B}'$. However, $\mathbf{B}' = 0$ and $\mathbf{N}' = -k\mathbf{T}$. Formula (5.9) now follows.

Chapter 6

6.7.4 If x and y are concentrations of the opposing forces, the equations are $x' = -ay$ and $y' = -bx$ with a and b as positive constants. Modifying the equation for x gives $x' = -ay - cx$, with $c > 0$. The sole equilibrium point for the unmodified x and y equations is the origin in the plane. The Jacobian for this linear system has a zero trace and negative determinant, showing that the origin is unstable, a so-called saddle point. Writing the equations in the form $dy/dx = bx/ay$ and separating variables as in Exercise 6.7.1 gives $a \int y \, dy = b \int x \, dx$, integrated from the initial values x_o and y_o. The result is $a(y^2(t) - y_o^2) = b(x^2(t) - x_o^2)$, which describes a family of hyperbolas $ay^2(t) - bx^2(t) = $ constant, where the constant is positive or negative depending on the value of $y_0 - (b/a)^{1/2}x_0$. In the positive quadrant (only positive concentrations are of interest) both x' and y' are negative. Therefore, depending on the positive initial values of the opposing forces and on the constants a and b, both x and y decrease in value with either x or y reaching a value of zero (namely, is annhililated) with the opponent remaining undefeated, except when the hyperbola goes through the origin. Typical trajectories are shown in the figure.

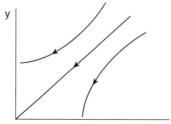

6.7.7 Differentiate $E_1' = a_1 E_2 - k_1 E_1$ to obtain $E_1'' = a_1(a_2 E_1 - k_2 E_2) - k_1 E_1'$. But $a_1 E_2 = E_1' + k_1 E_1$ from which we find that $E_1'' = (a_1 a_2 - k_1 k_2)E_1 - (k_1 + k_2)E_1'$. Now let $u = E_1'/E_1$. Then a simple computation shows that $u' = -u^2 - (k_1 + k_2)u - (k_1 k_2 - a_1 a_2)$. This first-order equation in u is known as a Riccati equation. Setting the right side of the equation to zero gives us the two equilibrium points, namely, $\lambda_1, \lambda_2 = -(k_1 + k_2)/2 \pm [(k_1 - k_2)^2/4 + a_1 a_2)]^{1/2}$. To obtain one positive and one negative root, as required for instability, it is necessary that $\theta = a_1 a_2/k_1 k_2 > 1$, replicating the condition obtained in Section 6.6. This is shown schematically here.

6.7.9 Inserting relations (6.16) into the equation for C in (6.15) gives $C' = C^2 - (Z_o + E^T + k_m)C + E^T Z_o = (C - r_1)(C - r_2)$, where r_1 and r_2 are the two roots of the quadratic right-hand side of the equation for C. A straight-forward computation shows that these roots are real and positive. Using the method of separating variables used in Exercise 6.7.1 the equations can be expressed as $\int dC/(C - r_1)(C - r_2) = \int dt$, and this can be integrated to yield $\log[(C - r_1)/(C - r_2)] = \log r + (r_1 - r_2)t$, where r is r_1/r_2. Solving, we obtain $C(t) = [(r_1 e^{-(r_1 + r_2)t} - r r_2)/(e^{-(r_1 + r_2)t} - r)] \to r_2$, as $t \to \infty$.

Chapter 7

7.6.2 If x jumps suddenly upward or downward it is a discontinuous function of E at the bifurcation point, which means that $G_x(x, E)$ must be zero there. Because x is also an equilibrium, $G(x, E)$ is, of course, zero as well.

7.6.3 The nullclines for the equations are exhibited in the figure here and we see that there is a positive equilibrium where the straight line $y = x$ meets the humped curve defined by $x' = 0$. Assuming $x > 0$ and letting $y = x$ in the right side of the equation for x gives a quadratic equation $x^2 + bx/r - 1 = 0$ for the equilibrium values of x. Choosing the positive root and substituting into the Jacobian matrix for the linearized system results in the matrix

$$\mathbf{A} = \begin{pmatrix} -rx + bx^2/(1 + x)^2 & -bx/(1 + x) \\ s & -s \end{pmatrix}$$

The determinant of $\mathbf{A}$ is readily computed to be positive s, while the trace of $\mathbf{A}$ is the expression $\mu = s - s^*$, where $s^* = bx^2/(1 + x)^2 - rx$. If b/r is large enough, s^* is positive and so μ can be positive or negative. With $\mu < 0$ the equilibrium is an attractor and it is a repeller for $\mu > 0$. For small enough μ the eigenvalues of $\mathbf{A}$ are complex because the trace is much smaller than the determinant. Also the derivative of the trace with respect to μ is 1. This suggests that $\mu = 0$ is a Hopf bifurcation point.

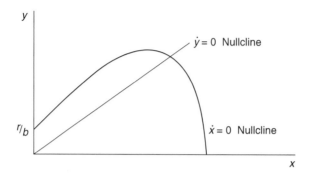

7.6.5 From linear algebra it is known that the trace and determinant of a matrix $\mathbf{A}$ are respectively the sum and product of the eigenvalues. Therefore the trace of the Jacobian $\mathbf{A}(s)$ is $2a(s)$ and its determinant is $a^2(s) + b^2(s)$. Moreover the derivative of the Jacobian with respect to s is simply $2a'(s)$. The result follows from Lemma 7.1 by setting s to zero.

Chapter 8

8.6.3 Equations (8.23) without the diffusion term have three equilibria: $(0, 0)$, $(0, 1)$, and $(1, 0)$. The Jacobian of the linearized system is

$$
\mathbf{A} = \begin{pmatrix} 1 - 2u - av & -au \\ -bv & 1 - 2v - bu \end{pmatrix}
$$

Only in the case of $u = 1$, $v = 0$ is the determinant of a positive and the trace negative (recall that $a < 1$ and $b > 1$).

8.6.4 The argument is identical except that $1 - a$ is replaced by $b - 1$ to give a minimum wave speed of $q \geq 2(b - 1)^{1/2}$.

Chapter 9

9.6.2 All integrals are over the region R in what follows. If $\int c(x)\,dx \leq 1$ then $p(x) \geq c(x)$ and $r \int [b(x)/a(x)]\,dx = \int c(x)\,dx \geq \int p(x)\,dx = 1$, whereas if $\int c(x) \geq 1$, then $p(x) \leq c(x)$ and $r \int [b(x)/a(x)]\,dx =$

$\int c(x)\,dx \geq \int p(x)\,dx = 1$. It follows that the dividing line between values of r for which $p(x) \geq c(x)$ and $p(x) < c(x)$ is $r^* = (\int [b(x)/a(x)]\,dx)^{-1}$.

9.6.3 The probability of contact P is $\int g(x)p(x)(1-h(x)\,dx$, with $\int g(x)\,dx = 1$ and $\int h(x)\,dx = d$. The patrol plane wishes to choose g to maximize P, whereas the submarine wants to pick h to minimize P. This problem is worked out in detail in reference [39] by Morse and Kimball.

9.6.7 The result is an immediate consequence of $G_1(u, v) = -G_2(u, v)$ combined with the minmax inequalities (9.22).

Index